P9-ELP-958

Managing Energy Costs: A Behavioral and Non-Technical Approach

Managing Energy Costs: A Behavioral and Non-Technical Approach

John Eggink

Library of Congress Cataloging-in-Publication Data

Eggink, John, 1959-
Managing energy costs : a behavioral and non-technical approach / John Eggink.
p. cm.
Includes bibliographical references and index.
ISBN 0-88173-544-2 (print) -- ISBN 0-88173-545-0 (electronic)
1. Energy conservation. 2. Energy consumption. 3. Power resources--Cost control. 4. Adjustment (Psychology) I. Title.
TJ163.3.E37 2006
658.2'6--dc22

2006041283

Published by The Fairmont Press, Inc.
700 Indian Trail
Lilburn, GA 30047
tel: 770-925-9388; fax: 770-381-9865
http://www.fairmontpress.com

Distributed by Taylor & Francis Ltd.
6000 Broken Sound Parkway NW, Suite 300
Boca Raton, FL 33487, USA
E-mail: orders@crcpress.com

Distributed by Taylor & Francis Ltd.
23-25 Blades Court
Deodar Road
London SW15 2NU, UK
E-mail: uk.tandf@thomsonpublishingservices.co.uk

Printed in the United States of America
10 9 8 7 6 5 4 3 2 1

0-88173-544-2 (The Fairmont Press, Inc.)
0-8493-8201-7 (Taylor & Francis Ltd.)

Contents

Acknowledgements

Much of the research sited in this book comes from government-sponsored programs. Material from the U.S. Department of Energy, the U.S. Environmental Protection Agency, Carbon Trust (funded by the U.K. government), the Canadian government's Department of Natural Resources, the State of California Flex Your Power program, and Lawrence Berkley National Laboratory contributed heavily to this book. Special thanks should also go to George Caraghiaur, Paul Gateo, Ted Klee, Philippe Delorme, Rick Stetler, Mark Feasel, William Rutter, Linda Eggink, Tatjana Grzenia-Eggink, and Cliff Eggink for numerous suggestions, input or motivation. Of course, none of this would have been possible without the unconditional support of my wife, Danielle, and our two daughters, Ellie and Maggie.

Introduction

...if you don't turn out the light, he takes the lights out of the ceiling.

—Maria Shriver discussing Arnold Schwarzenegger's home energy policy with Oprah

Have you ever heard the phrases "Turn off the lights" and "Shut the door?" Slogans like those echoed throughout my childhood. Later in life, as a member of the workforce, I was amazed that I could leave lights and equipment on without hearing about my wastefulness. Was this an implicit directive that waste was okay? Surely not, as one company I worked for rationed pencils. Yet, many energy-consuming devices were on needlessly. How could this be? Why was electrical waste different, not subject to the same cost controls and oversight as other items? At some companies, it appeared that electrical waste was almost encouraged, squandering energy was a sign of success; more usage was deemed better, and somehow reassuring.

This unchecked electrical waste seemed odd to me; so I probed deeper, questioning the status quo. The initial answers were surprising, ranging from pure apathy to myths, such that it is cheaper to leave lights on than to turn them off. I wondered, were these myths true? What did this waste cost? Were we talking a few cents or millions of dollars?

Having an engineering degree, I knew I could tackle any technical issues. Researching this topic became a hobby, or perhaps a quest. I interviewed energy managers, engineers, and executives. I spent many nights and weekends at libraries digging up what little research existed. Technical journals had largely overlooked the role of human behavior in energy usage. Thus, it took many years to develop a comprehensive understanding of the behavioral aspects of energy waste.

I found that "leaving devices on unnecessarily" was a widespread, pervasive, and acknowledged problem costing businesses billions of dollars each year. Most organizations fail to adopt simple low cost, or no cost, energy saving activities. Roughly 15 percent, or more, of an electric bill can be attributable to this oversight. This happens in all types of businesses: production plants that remain partially powered-up over weekends, and offices that keep computers on all night needlessly. These same businesses that measure chemical usage to the drop, track metal shavings to the ounce, and count pencils do not effectively manage the human component of energy management. On the other hand, companies that do address the behavioral aspects of energy consumption reap substantial benefits. IBM, for example, estimates that it saved $17.8 million one year by encouraging employees to turn off equipment after completing tasks and to moderate their use of lighting.[1] Verizon spent less than $5,000 training employees to save energy and reduced its California electric bill by $750,000.[2]

Many companies save hundreds of thousands of dollars on their monthly electric bills, simply through effective employee training and workforce management techniques. Typically, this methodology is called an energy awareness program. The initial goal of the program is to increase an organization's energy IQ, or awareness of energy consumption, its cost and environmental consequences, and to make everyone accountable for energy consumption. The training encourages personnel to turn off devices, and to look for other opportunities to reduce energy costs, such as reporting equipment or maintenance issues, rescheduling loads, negotiating better tariffs, and a host of other low-cost or no-cost actions that can substantially reduce energy costs. Our individual tendency to leave devices energized long after we are through using them can be significantly reduced through effective management and training.

Becoming "energy aware" and sufficiently knowledgeable about electrical energy consumption does not require a lot of time. Everyone in an organization can read this book quickly, in a few hours or less, and gain common understanding of behavioral energy issues. This minor time investment can result in substantial energy savings. Energy awareness programs include proven, repeatable, and affordable measures that reduce energy expenses. Any organization that truly wants or needs to reduce energy costs or reduce dangerous emissions can do so with an energy awareness program. The cost, thereof, is typically low. The investment payback period is short, usually within the same fiscal year.

Ideally, a senior manager gave you this book to solicit your support in reducing energy. If not, please pass it "up" to an executive. Managers are hard pressed for time, so don't be discouraged if they don't immediately jump on the bandwagon. In the meantime, if you want to reduce energy expenses, start a grass roots movement in your sphere of influence.

References

1. John Douglas, "The Energy Efficient Office," *Electric Power Research Institute EPRI Journal*, July/August 1994, P. 16.
2. California Flex Your Power Initiative, *Business Guide 3: Target Business Employees for Energy Conservation in the Workplace*, p 9 http://www.fypower.org/pdf/BPG_Biz3_Target_Employ.pdf (June 22, 2006).

Section I

Linking Behavior and Energy Consumption

Chapter 1

Unique Characteristics of Electricity that Increase Energy Costs

Verizon's employee awareness efforts cost less than $5,000 to implement—but yielded savings of $750,000 and 10 million kWh in California alone.

—California Flex Your Power Program

For all the corporate urging to cut costs however possible, businesses continue to waste energy at an amazing rate. Some of this waste is attributable to the unique physical qualities of electricity. For example, our inability to visualize the flow of electrical energy, or our inability to stockpile and store electrical energy are unusual traits for a commodity. These two attributes alone greatly differentiate electrical energy from most other commodities. These unique qualities govern the pricing, procurement, delivery, and consumption of electrical energy, and influence our propensity to waste electrical energy.

Businesses that have a thorough awareness of the key distinctions between electrical energy and other commonly used commodities are better prepared to reduce energy costs. We will now examine some "cost increasing" attributes of electricity. This is a necessary and painless step towards reducing energy expenses through energy awareness.

ATTRIBUTE #1: ELECTRICITY IS INVISIBLE

Businesses typically measure material waste. We can easily see the pile of scrap or count the number of rejected parts. We can see and hear a running water faucet. Whether it is in gallons or number of pieces, we can visually quantify waste, not so with electricity; therein, lies one problem.

Other than the occasional lightning strike or spark, we cannot see electricity. This invisible characteristic makes it easier to waste electrical energy. For example, if we tip over a carton of milk, we can see the milk pour out. Since we don't want a bigger mess, we usually grab the carton and stop the spill. That's human nature. We can see the puddle, peer in and examine the remaining amount of milk. We can visualize the loss. Because electrical flow is invisible, there is no mess to clean up, no obvious way to visualize the loss, and no reflex to stop the spill.

Consequently, depending on the size of the facility, there may be hundreds or perhaps thousands of electrical "leaks." That is to say, there are many devices needlessly consuming energy. They may be small devices such as desktop computers, or larger devices such as whole building air conditioning units. The point is, unlike spilt milk, we can't naturally see the accumulated energy lost.

Fortunately, we can measure the amount of energy consumed, the amount of money paid for it, and the emissions left behind. Disseminating this energy information is an excellent way to counteract the invisible nature of electrical energy. A good example of this comes from Harvard University. While entering the dinning hall at the Spangler Center I could not help but notice a poster board resting on a tripod. I have duplicated this poster at the top of the following page.

This low-cost poster accomplishes several things: (1) it shows a correlation between energy conservation and financial savings; (2) it appeals to environmentally concerned individuals by emphasizing the environmental aspects of energy conservation; (3) it equates invisible energy to tangible items such as barrels of oil

> If HBS [Harvard Business School] can achieve a 3% total energy reduction in one month it would save $8,967.93 and offset 105,052.89 pounds of harmful CO_2. This amount of reduction is equivalent to saving 111 barrels of oil or 5,427 gallons of gasoline.
>
> | Facility | Savings |
> |---|---|
> | Dillon | 15% |
> | Loeb | 15% |
> | Cotting | 10% |
>
> 1st Annual Energy Competition 10, July, 2006 – 10, August, 2006
> HBS GREEN TEAM (and contact information)

and gallons of gas; (4) it creates energy awareness among the students, staff, and guests; (5) it communicates that saving energy is important and results are monitored and measured; (6) it serves as the scoreboard for an energy saving competition between several campus buildings. All the contestants receive feedback on their efforts. The winners, tied at 15% savings, are recognized for their accomplishments. (I did omit the names of five buildings tied for last place.) What's more, the "percentage based" competitive measurements utilized by Harvard are, in this case, advantageous. It is easier to visualize reducing energy consumption by 1%, than reducing a fixed (absolute) metric such as $3,000 dollars or 300 kilowatt-hours. Individuals are more motivated by relative values than absolutes.[1] Please note, the terms "savings," "offset" and "avoided costs" are often interchangeable when discussing energy conservation topics.

Another way of making energy more visible is to graph the relevant energy metrics. Graphs can provide a "picture" of the energy usage. Our eyes are good at spotting patterns and quickly identify anomalies, or opportunities to save energy. I can't emphasis enough the importance of properly graphing interval energy data versus just utilizing rows of figures. It is impractical to study thousands, or often millions of lines of data. However, graphically

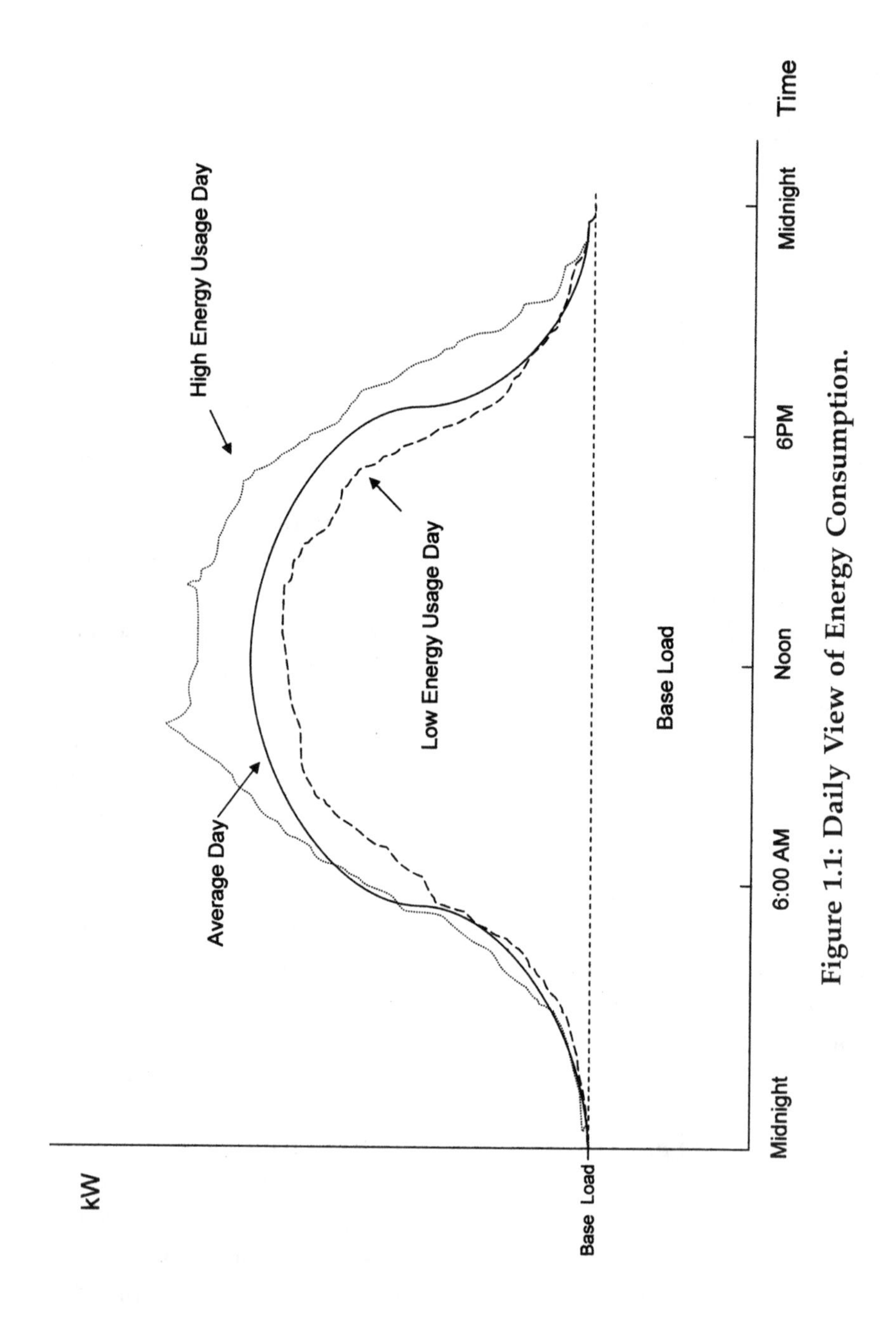

Figure 1.1: Daily View of Energy Consumption.

displaying the data enables the eye to spot deviations quickly, which turns the data into useful information.

Figure 1.1 shows a typical office building "load profile." A load profile is a graph of the variation of electrical consumption versus time. The vertical, or y-axis, represents the amount of electrical power consumed. Since the graph's purpose is to show the consumption pattern the y-axis does not indicate the actual numerical amount. The middle line represents the average electrical load. The outer lines represent the hypothetical minimum and maximum values. Weather, building occupancy, manufacturing output, or energy saving activities can significantly affect the load profile.

Figure 1.2 shows a weekly load profile. We can see the energy consumption increase throughout the day and drop every night. Notice that on Wednesday night the power use did not drop as expected. There could be several reasons for this. One possibility is energy usage from a special event, such as a concert or dinner banquet. Another possibility is the night cleaning crew failed to turn off equipment and lights. Another concern is the high weekend energy use. Each building will be different, hence the need for human intervention to analyze the graphs.

Figure 1.3 shows a scatter plot of energy usage. Each box in the graph represents an individual day. Moving from a box horizontally to the y-axis shows the amount of energy used that day and the x-axis is shows the outside air temperature. The x-axis can be any energy influencing factor, such as the production rate in a manufacturing facility. In an office building, the outside air temperature can greatly affect energy consumption.

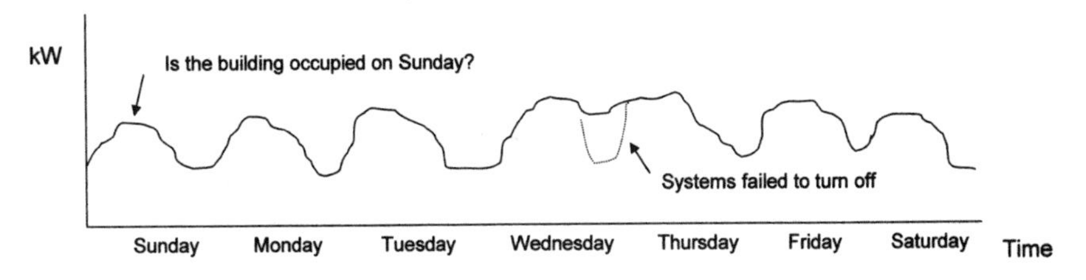

Figure 1.2: Weekly View of Energy Consumption.

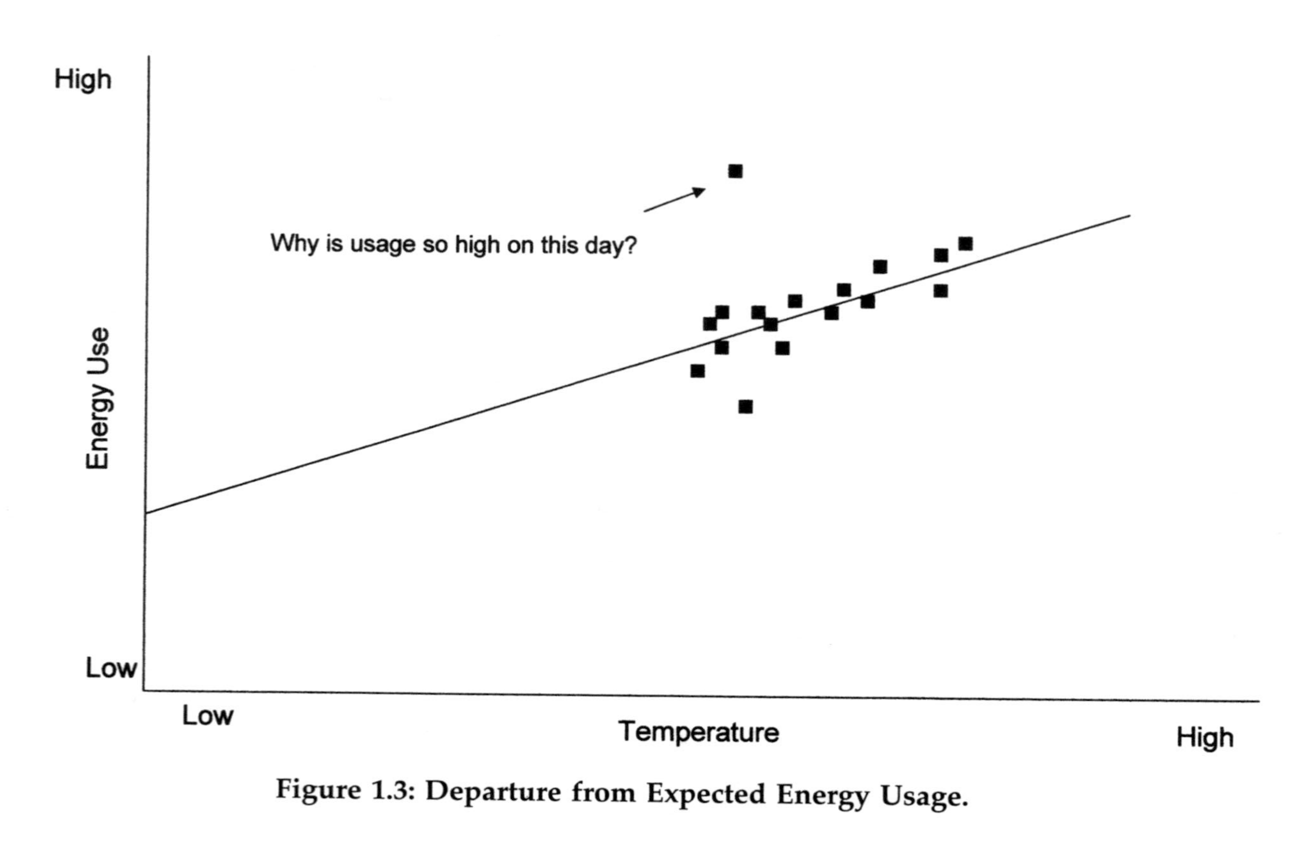

Figure 1.3: Departure from Expected Energy Usage.

In a manufacturing facility, it is often the number of widgets, or type of widgets produced that influence energy consumption. The solid line is the forecasted load. As expected, the energy use increases with the temperature, however we can easily see there is one day that greatly exceeds the forecasted energy usage. A graph like this allows any non-technical person to spot patterns in energy consumption quickly and easily. Graphs like these make the energy usage visible and understandable. It is also important to generate the graphs close in time to the actual usage. Rapid feedback is critical for efficient learning and quick resolution of high-energy events.

ATTRIBUTE #2 ELECTRICAL ENERGY IS THE ONLY PRODUCT CONSUMED THE INSTANT IT IS MADE

The courts are split on whether electricity is considered a product (good) or a service. The answer depends, in part, on where you live, and can be a decisive factor in litigation. Regardless of the legal definition, electrical energy is consumed the instant it is generated. Let's think about this for a second. Virtually all other products we consume can be manufactured in advance and counted as inventory. The electrical energy we use at work and home arrives at near the speed of light from a utility generator.

One consequence of not being able to effectively store electrical energy, or keep any electricity inventory, is that utilities must always have the capacity to manufacture the maximum required electrical energy that customers demand.[2] If the consumers demand for electricity is greater than utilities can supply, it creates tremendous problems and can cause the entire grid to collapse. One analogy is perhaps that of riding a bike. It is easy to balance and stay up when we are moving quickly, but as the bike slows, we become unstable. If the utility generators cannot maintain their desired speed, corresponding voltage, and frequency, the system will crash. To avoid a catastrophic failure, utilities will turn off electricity to small geographical areas to keep the generators running. This controlled outage, often called a rolling blackout, is a desperate maneuver designed to keep the power grid from collapsing. Power outages can have tragic consequences and even result in loss of life, as traffic signals and other essential equipment fail.

Building the necessary power plants and transmission lines to handle this maximum electricity use is an expensive endeavor, especially, if it is only needed a few short times during the year, usually on the hottest summer days, when air conditioning use is at its peak. So ideally, utilities try to encourage their customers not to consume electrical energy during the peak usage periods in an attempt to reduce the maximum demand for electrical power. How does the utility do this? Simple, by adding additional charges to make electricity more expensive during the peak time of day when the electrical energy consumption at its highest. The higher prices discourage electrical energy consumption during peak times, or conversely, the higher prices help finance the additional infrastructure.

The maximum demand period varies by region, but generally falls between noon and 7:00 P.M. Depending on the particular utility tariff, energy costs can be substantially higher during certain peak times. We should especially watch our usage when rates are the highest, typically from lunchtime though the early evening when the actual peak demands typically occur. This can be as simple as turning off computers and lights when we leave for lunch or home. Industries can sometimes reduce electrical bills by staggering lunches. For example, if warehouse operators recharge forklift batteries during their lunch break, the simultaneous charging of all the forklifts can create a peak demand and larger than necessary electric bills. Staggering the forklift operator's lunch schedules reduces the instantaneous electrical power and lowers the electric bill.

One extremely positive ramification of immediate consumption is that conservation results are instantaneous. If we get up now and turn off one light, one copy machine, one assembly line, or any appliance, instantly the utility generator transfers less electrical energy, and the electric bill is reduced.

ATTRIBUTE #3: ENERGY MANAGEMENT IS DIFFERENT FROM ENERGY CONSERVATION

At this point, you may be thinking that shifting energy use to different times of day is not really reducing energy consumption. This is counter-intuitive to some people, but an energy management initiative could actually increase the total amount of energy consumed, but still reduce the total price paid for energy. Cost reduction may be achieved by shifting loads to times when energy is cheaper, thereby receiving smaller utility bills, and yet use the same or even more energy.

Energy managers may negotiate with energy suppliers and select the cheapest utility tariffs. The energy manager may contact the water company to request that evaporated cooling tower water (water used by the cooling process and not returned to the sewer system) not be subject to sewer charges, thereby reducing the utility bill. Sounds simple, but many organizations do not bother to do this. Managing the cost and reliability of energy is just as important as managing energy consumption.

ATTRIBUTE #4: ELECTRICAL TERMINOLOGY IS COMPLICATED

The technical aspects of electricity can bore and confuse people. Aside from being intangible, electrical units are often convoluted. If a gasoline station sold gas the same way electric utilities sold electricity, gas stations would charge us for "kilogallons-pumped-per-second per hour" in lieu of gallons.[3] It does not need to be that hard.

Executives, plant or facility managers, and engineering staff should know three main billing terms. They are kilowatt-hour, watt, and power factor. Simple definitions are as follows:

- Kilowatt-hour (kWh)—A kilowatt-hour is the basic unit used to measure and sell electrical energy. It is often called energy consumption, or sometimes just electrical energy. Think of a kWh as you would an odometer on a car; it is a cumulative sum. The k in kWh represents the number 1,000.

  The kilowatt-hour charge can be broken down into kilowatt-hours consumed by time period. For example, the utility may charge one rate for kilowatt-hours consumed "on-peak" (for example, 10:00 A.M to 7:00 P.M weekdays and holidays) and a different rate for kilowatt-hours consumed "off-peak" (all other hours). This is called "time-of-use" (TOU) billing.

- Watt (W)—The other frequently used term is the watt. It is a measure of energy flow often referred to as power, active power, demand, kW (1,000W), peak demand, kW demand, or megawatt (1,000,000 W). One reason utility companies call this demand is that customers just take it. Few people call and say, "Can I plug in a few more devices?" We expect a limitless supply to be available. Think of watts as the vehicle equivalent of a speedometer; it is an instantaneous reading. It represents the amount of electrical energy consumed at any given moment.

  Utilities often add an additional charge for kW demand. In this case, kW demand is an average value of watts, typically averaged over 15 to 30 minutes. There are variations on how utilities calculate the average demand, but the same principles hold. If we go back to our speedometer analogy, remember that kW is an instantaneous value equivalent to our speed, say 60 miles an hour. If you where to drive for one hour your speed may vary, it may be 0 miles an hour for a few minutes then rise to 70 miles an hour for a few minutes. To calculate the average speed, you simply take the number of miles you drove and divide by the length of time you drove. So, if you drove 59 miles in one hour you averaged 59 miles an hour. This is the equivalent of what utilities do. They take the kWh used over an amount of time and divide by the same time period and get an average kW (the "time component" cancels out on the numerator and denominator). For example, 2,000 kWh used in 15 minutes (1/4 hour or .25 hours) yields a demand of 8,000 kW (2,000kWh/.25h = 8,000 kW). The highest demand reading in a billing period becomes the "peak" demand.

Utilities bill the "peak demand" to recoup the costs of building and supporting the infrastructure required to deliver the maximum kW needed, even if only used once or twice a year. In some parts of the country, demand charges may account for roughly half the electric bill, so understanding how fast a facility uses energy may be just as important as knowing how much energy is consumed.

Usually, the demand charge is reset each month. That is, each month is billed for the peak demand that occurred in the month. There is a costly variation on the demand penalty called a Ratchet. A ratchet uses the higher of the current peak demand or any past peak demand. With a ratchet tariff in effect, setting a new peak demand level has consequences well into the future.

- Power Factor—Motors and transformers utilize electrical current to create magnetic fields. In addition to the power that actually produces useful work, utilities must also supply the power to create these magnetic fields. Further technical details are beyond the scope of this book, but a simply analogy is the foam in a glass of beer or soda. When pouring the beverage, foam can accumulate in the glass. The foam causes two problems: one, it takes up space in the glass; two, it is an inefficient use of the beverage. Power factor represents the electrical equivalent of foam, which in reality is the energy used to create magnetic fields. Power factor is always a dimensionless number between 0 and 1.

  Many utilities do not charge for poor power factor, but those that do typically start charging additional fees at a power factor of less than 0.95. There are some no cost ways to reduce power factor such as minimizing the operation of lightly loaded or idling motors. If the power factor is still out of tolerance, there are power factor correction devices that can mitigate the problem.

When dealing with a larger audience, rather than use technical terms such as kilowatt-hours or watts, it is sometimes preferable to express energy in simple and easy to understand units such as dollars or percent to goal. These units are often more familiar and tangible. A good example of this comes from "Climate: Making Sense and Making Money:"

> At a large hard-disk drive factory, the cleanroom operator started saving lots of money once the gauge that showed when to change dirty filters was marked not just in green and red zones, but in "cents per drive" and "thousand dollars' profit per year.[4]

Again, when measuring, tracking, or reporting energy usage, it is preferable to express units in terms that are understandable and measurable. Sometimes this requires monetizing the energy usage and discussing it in terms of its financial cost. Sometimes this requires converting the energy units to emissions equivalents, or as a percent to goal, or compared to prior usage. Expressing energy consumption in units of pollution such as carbon dioxide (CO_2), nitrogen oxides (NO_x), or other volatile organic compounds and particulates can be effective in decreasing energy costs. Organizations such as universities or hotels should place extra emphasis on the emission equivalent of energy consumption, such as pounds of carbon dioxide. We will discuss this more in later chapters.

Information should be timely. An electric bill showing abnormally high-energy usage a month ago is not as useful as a daily report that enables immediate cost reduction activity. The main concept is that the metrics need to be relevant, personal, accurate, timely, and easy to understand.

ATTRIBUTE #5: ELECTRICITY IS THE CARRIER OF ENERGY, AND NOT AN ENERGY SOURCE

We often forget that electricity is merely the mechanism used to transport energy. Electricity is a secondary energy carrier and not an energy source. Its availability requires conversion from other energy sources or carriers, mainly from:[5]

- Mechanical power derived from thermal energy, or heat. The heat comes from burning fossil fuels such as coal, natural gas, and petroleum, or in lieu of burning fossil fuels, a nuclear reaction can generate heat.

- Directly harvested mechanical power, such as wind or hydroelectric, in which wind or water turns a generator transferring energy.

- Light (solar radiation) such as solar cells.

- Chemical energy, such as fuel cells, that turn natural gas into electricity.

Sitting in our modern offices, we no longer experience a direct correlation to our dependence with the actual energy sources. We cannot see the fuel that is burning to create the energy used to power our lights and equipment, but it is burning. The majority of electrical power comes from nonrenewable resources such as coal, oil, and natural gas (see Figures 1.4-1.5). In the United States about 50 percent of electrical energy comes from coal. Worldwide, burning coal generates 40 percent of electrical energy.

The fossil fuel burned by the utility to power the unneeded lamp is lost forever. Wasting energy devours excessive amounts of these finite resources. Reconnecting with the reality that electrical energy comes primarily from burning a finite supply of fossil fuels can increase our incentive to conserve.

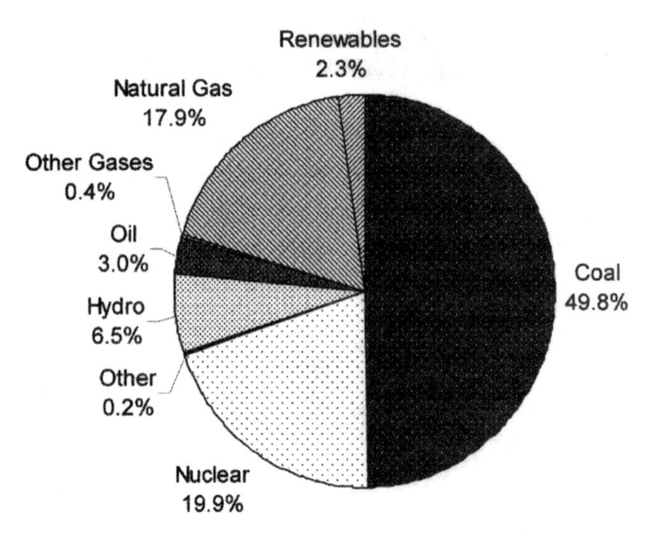

Figure 1.4: United States Electrical Energy Sources. ***Source data from U.S. Energy Informaton Administration (EIA), 2004 data.***

ATTRIBUTE #6: ELECTRICAL ENERGY IS SUPPLIED EFFORTLESSLY

Not long ago, during the winter season, our ancestors had one of two choices. They could build a small fire and sit close to keep warm, or build a large fire and sit further back. Either way the fire provided the necessary heat to keep them alive. There was a direct consequence to building a larger fire; it required considerably more effort to gather the extra wood.

Our world is different. No such obvious choices are required. Utility companies deliver energy straight to our homes and businesses on a continuous basis. We do not have, nor associate, any inconvenience with using energy delivered real-time on demand, nor do we feel any need to ration a resource that seems limitless. The long lag time between usage and receiving the utility bill further robs us of helpful feedback on our energy habits.

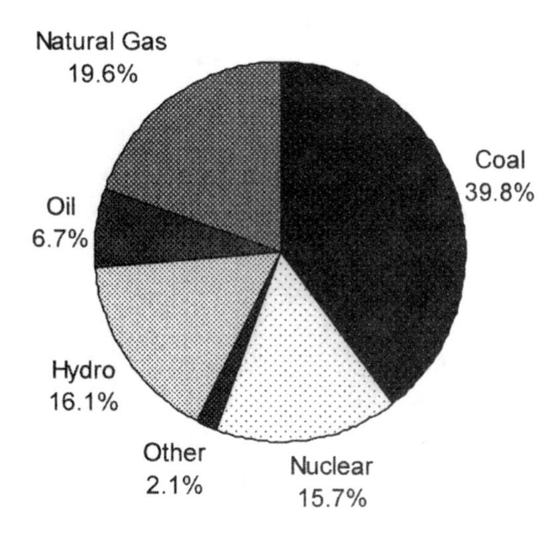

Figure 1.5: World electrical Energy Sources, ***Source data from the International Energy Agency (IEA), 2004 data.***

SUMMARY

Some electrical waste is attributable to unique attributes that separate electrical energy from other commodities. Understanding these qualities is a good step towards reducing energy expenses. They are:

- Electrical energy is invisible, but its associated costs and pollutants are not.
- Electrical energy is the only product consumed the instant it is made. We need to be conscious of not only how much energy we use, but also when we use it.
- Energy management is different from energy conservation. Energy management focuses primarily on reducing energy costs.
- Technical jargon can interfere with energy saving initiatives. If possible, express energy consumption is dollars, emissions,

or percent to goal. When dealing with a smaller or more manageable set of individuals ensure that everyone understands the basic energy billing units such as kWh and kW. Simple definitions are:

— A kilowatt-hour (kWh) is often referred to as electrical energy, consumption, or energy usage. It is the basic unit in which electrical energy is bought and sold. Think of a kWh as you would an odometer in a car; it is a cumulative sum.

— A watt (w) is the rate electrical energy is used and often called electrical power. Think of watts as the vehicle equivalent of a speedometer; it is an instantaneous reading. The average rate of power flow over a period of time is "demand" kW.

- Electricity is a unique carrier of energy, not an energy source. Typically, utilities burn fossil fuels to supply our electrical energy.

- Utilities supply electricity effortlessly and abundantly on demand. This makes it easy and effortless to waste electrical energy.

References

1. Daniel Gilbert, *Stumbling on Happiness* (New York: Random House, 2006) P. 138-141.
2. I should point out that we can store electrical energy in a battery, but batteries cannot power large loads such as homes or businesses on a long-term basis. Batteries are best suited for smaller portable appliances, and devices that require temporary power.
3. I have yet to discover why utilities bill in the counter intuitive unit of kilowatt-hours. For some people, mainly those that pay attention to language or understand the units, the kilowatt-hour can be confusing. The most common issue is the "hour" component in "kilowatt-hour" it raises the question: does this mean that we need to use a kilowatt-hour for an hour? Not really, we used 1,000 watts for an hour. The kilowatt-hour is a unit of energy equal to that expended by one kilowatt over one hour. It would probably be simpler to bill in a basic unit of energy such as a Joule. A watt is a joule/sec. Hence 1kWh = 1,000 Wh =1000 * Joule/sec H * 3600sec/H. Cross out the Seconds and we get 3,600,000 Joule or 3.6M Joule = 1kWh. Alternately, GJ = 277.8 kWh.
4. Amory B. Lovins and L. Hunter Lovins, "Climate: Making Sense and Making Money," November 13, 1997.
5. Aviel Verbruggen, "Stimulating Electricity Efficiency," *Slovenski E-Forum. Changing Paradigms in the Energy Sector,* November 13-14, 2003. P. 3.

Chapter 2

How Attitudes Influence Energy Consumption

> ***Those of us who call ourselves energy analyst have made a mistake... we have analyzed energy. We should have analyzed human behavior.***
>
> *—Lee Schipper*
> *Energy Economist*[1]

People control thermostats, light switches, computers, process machinery, and a host of sophisticated devices. To manage energy effectively, we need to accept that our individual behavior can have a profound effect on the amount of energy these devices consume, both at home and in the workplace. Studies show that the proper operation of energy consuming devices can produce dramatic energy savings, often 15% or more of the current energy bill.[2] Yet, most organizations fail to capitalize on these saving opportunities.

Understanding human mind-sets, our habitual attitudes and inclinations that determine how we interpret and respond to situations that influence energy consumption is a necessary step to maximizing energy savings. There are eleven common mind-sets covered in this chapter. It is fitting to start with our "resistance to coercion," since at this point most type A executives are thinking, "I'll just order my employees to save energy." If that worked, all would be well.

MIND-SET #1: RESISTANCE TO COERCION

Strong unilateral directives such as "turn of the lights," or "stop wasting energy" can be counter productive. There is a natural resistance to forced energy change. Some of us just don't like being told what to do. If we feel forced, or coerced, into saving energy, there can be a tendency to rebel. Appeals to reduce energy use might even have the reverse effect of increasing the desire for consumption.[3] That's right, ordering employees to save energy, without sufficient explanations, may actually backfire and increase energy consumption. This means that in most organizations, simply ordering people to turn off the lights or report energy inefficiencies can be extraordinarily ineffective.

However, inducing people to change energy habits through education, rather than coercion, is extremely effective. The education process needs to humanize the problem and provide positive reasons to conserve energy. When we reflect on positive aspects and success, rather than threats or coercion, appropriate behavior tends to be reinforced and we feel more in control of our circumstances and ourselves.

Each call to reduce consumption should provide knowledge and or be associated with a positive benefit to the action. The WWII "loose lips sink ships" campaign is an example of motivating the right way. The slogan doesn't just dictate what to do. It includes the "what's in it for me" motivator—I may die, or other people may die, if I do not follow security rules. Similarly, putting up a "turn off the lights sign" may not be effective, whereas an "electric spills raise the bills, please turn off the lights" sign may invoke a better response.

There is one clarification to the discussion above, perhaps a paradox. Coercive adaptation that comes from a government or an official regulatory body in the form of a law or mandate can be an extremely effective motivator. As much as we dislike orders, we recognize the need to comply with various laws, mandates, and emergency situations. In this case, the mandate from a government body becomes the explanation that can make the managerial edict successful. If we should find ourselves in a situation where energy reduction is mandated, or there is a temporary crisis, communicate the official dictates to all members of the organization immediately.

In times of crisis, the government agency may often appeal directly with the public using the mass media to explain the energy crisis and encourage conservation. California summoned comedians and used humor to encourage action. Japan has used movie stars. New Zealand, Norway and Brazil used pictures of empty reservoirs to help people visualize their energy crisis.[4] Some examples of the no cost, low cost, actions requested are:

- Consumers were asked to raise room temperatures a few degrees and re-set the timers on their pool pumps after an Arizona utility experienced equipment problems. Both measures cut peak power demand and avoided blackouts for several weeks until the faulty equipment was replaced.[5]

- Sweden successfully encouraged its consumers to briefly lower thermostats and to postpone non-essential electricity-consuming activities to cope with a day-long shortage one winter.[6]

- Millions of Brazilians unplugged their freezers during an electrical crisis. This was enough to achieve the 20% savings required by the government, at threat of power disconnection for the consumer. Interestingly, the energy consumption of freezers has yet to return pre-shortage levels.[7]

MIND-SET #2: OVER-ZEALOUSNESS BACKFIRES

You may be familiar with the pendulum effect when an organization adopts a new process. They may go overboard enforcing new activities, trying to swing the pendulum from one end of the scale to other within the blink of an eye. Let me give you an absurd example of what not to do: Go from a scenario where lights needlessly burn all day to a scenario where we are stumbling around in the dark and adding additional security patrols in unlit parking garages for safety. Even though it is well intentioned, we must guard against over-zealous conservation activities that can diminish long-term effectiveness, or are economically undesirable.

The goal we are striving for is to make our organization more competitive and more profitable through the efficient use of energy. We must distinguish between what is sacred and what is not. In an upscale retail store, for example, studies indicate effective display lighting can increase sales. The quest for higher sales justifies the extra energy required for brilliant lighting; this is not wasting energy as the energy is used productively. We should not sit in dark offices or reduce lighting levels below an optimum performance level. Raising the indoor air temperature from a comfortable 72 degrees Fahrenheit (22.2 °C) to 80 degrees (26.7 °C) is bound to affect the productivity of office workers wearing coats and ties. There are countless studies indicating the generous use of electricity helps businesses generate more revenue. For instance, studies have shown that:

- Proper and abundant lighting in offices and factory floors can increase work quality, increase productivity, and decrease absenteeism.

- High lighting levels can improve safety and security, thereby reducing crime such as vandalism, break-ins, or assaults.

- Bright lighting can decrease vehicle and pedestrian accidents.

- Effective lighting can increase sales by attracting more people to shopping centers and retail stores.

- Low lighting levels in offices can cause depression and negatively effect morale.

- Diseased meats and food can pass inspection under improper lighting.

Recognize that keeping lights off, when they're needed, is often more harmful than keeping them on, when they are not needed. Our individual physical well-being and the economic health of nations do depend on the consumption of electrical energy.

Do not feel guilty, or make other people feel guilty about using energy productively. We just need to increase our energy awareness and avoid wasting energy needlessly. There are plenty of opportunities to reduce energy costs by eliminating nonproductive energy use.

MIND-SET #3: BIAS AGAINST CONSERVATION

Some people see the preservation and careful management of natural resources as a sign of personal virtue, but not a personal necessity. If our bias is against environmental conservation, or if we are perhaps ambivalent towards conservation, as employees we should understand that conservation activities can save our company substantial amounts of money. Companies that neglect to manage costs do not fair well in the long run, and energy costs are frequently unmanaged. Thus, even if our company's environmental stance is stronger than our personal beliefs, we should strive to honor and enforce the company's position. What's good for the company's bottom line in the end, is good for our personal bottom line.

Additionally, corporate responsibility buys goodwill from investors and customers. Investors can easily track the performance of companies that manage environmental and social issues with Dow Jones Sustainability Indexes[8] or through magazines such as *Corporate Knights*[9] in Canada that annually rank "responsible companies." It is tempting to dismiss this as a feel good or soft issue, where a small fringe group of consumers and investors support personal causes. But that may be a mistake. There is a well-documented theory that companies that effectively manage energy have better than average management teams.[10] The better the management team, the better the companies performance will be versus its competitors. Jean Frijns, the chief investment officer of ABP Netherlands, sums it up well, "There is a growing body of evidence that companies which manage environmental, social, and governance risks most effectively tend to deliver better risk-adjusted financial performance than their industry peers. Moreover, all three of these sets of issues are likely to have an even greater impact on companies' competitiveness and financial performance in the future."

MIND-SET #4: THIS PROBLEM IS TOO LARGE FOR INDIVIDUALS TO SOLVE

We may ask ourselves: "Can my actions really make a dent in this problem? I'm just one person. Am I sacrificing my luxuries for someone else's benefit? How big a difference can I make by turning off a light?" These are great questions. Changing our behavior cannot solve a problem that requires universal support. If forced to act alone, we are apt to give up and leave energy management for governments, utilities, or industry, thinking it is an issue we cannot solve.

However, there is another way to look at it, interconnectedness gives every person power. We are all part of a vast array of interconnected people and organizations. This offers everyone the opportunity to act like leaders, to influence others, and make a big difference.

Never believe that a few caring people can't change the world. For, indeed, that's all who ever have.

—Margaret Mead

Realize that single individuals can make a difference. An American Express facility experienced $700,000 in energy savings after one employee contacted "ExpressLine" and asked whether turning off devices was a worthwhile endeavor. The American Express Environmental Protection Committee then researched whether employees could make a difference by diligently shutting off personal computers. The answer was, of course, yes. American Express then teamed with the local utility company, Con Edison, to launch an informational campaign to encourage employees to turn off devices. In this case, one person was the catalyst for the savings program at America Express. Thus, experience shows that anyone can make a difference.[11]

Two tips to leverage interconnectedness:

1. Create and form energy conservation teams to motivate other people to save energy. There is substantial evidence that energy teams operating as cohesive groups are very efficient at conserving energy and changing individual behavior. The best teams consist of people in local groups, often with varied job descriptions. Teams are easily motivated to reduce energy consumption and even enjoy healthy, good spirited, competition against each other. A person's desire to conform within the group is generally stronger than their unwillingness to conserve energy. Peer pressure is a strong motivator and is more effective than top down edicts. We can also expect less resistance, if forced conservation becomes temporarily necessary, such as in the historic California power alerts.

2. Obtain senior management "buy-in" for energy conservation; this provides additional encouragement for other employees to contribute and puts the change agents at ease. It is important that senior management back-up the verbal commitment with action. One word of caution: Employees can determine if the executive commitment is genuine. One quality manager, to whom I spoke, explained that whenever he asked employees to reduce electrical waste, they in turn would fire back: "Then why are the parking lot lights [across the entire campus facility] on in the daytime?" Therefore, in the mind of the employees, the senior management had obviously seen these lights on, but did nothing. Clearly, then, senior management did not really care about energy conservation. By the way, depending on the types of lights and the local cost of energy, it typically costs

from $30 to $50 to light a single parking space for one year. Lighting just 500 parking spaces needlessly during the day can increase the annual electric bill by approximately $25,000 a year.

MIND-SET #5: BELIEF THAT TECHNOLOGY WILL SAVE THE DAY

Taking responsibility for energy consumption is a critical success factor towards significant energy reductions. When discussing that coal, oil, and natural gas supplies are finite and non-renewable, or that fossil fuels create harmful emissions, or that energy costs are too high, someone will usually say "no worries, new technologies will save the day." Similar to the number one grossing movie of 1985, "Back to the Future," many of us are optimists, thinking we will drop empty beer cans in toasters and power our homes and cars. A surprisingly large number of people feel this way. In one survey, 72% of the respondents felt technology would find a way to solve our energy problems.[12] In fairness, there are many promising technologies such as:

- Biofuel processing factories, which make fuel from biological sources, such as plants and waste material.

- Fuel cells, which convert hydrogen or other fuels into electricity cleanly.

- Solar cells, that turn sunlight into electricity.

- Wind turbines, which convert wind energy into electricity.

- Safer, more efficient, nuclear power plant technology (nuclear plants do not produce CO_2 emissions), though some power plants can require up to 200 tons of raw uranium a year and the industry continues to suffer image problems as a result of the Three Mile Island and Chernobyl disasters in 1979, and 1986. Concern about the storage of the nuclear waste is also an unresolved issue.

As promising as these alternate sources of energy are, they are not yet poised to eliminate our fossil fuel dependence, which is likely to continue for the foreseeable future.

Despite our optimistic tendencies, we should stop and consider, if hoping for a *Back to the Future* scenario is an appropriate strategy. In 1955 a vacuum cleaner company forecasted in *The New York Times*: "Nuclear-powered vacuum cleaners will probably be a reality in ten years."[13] Four years later, the U.S. Postmaster General predicted, "We stand on the threshold of rocket mail."[14] Clearly, energy-producing technologies have lagged behind some optimistic schedules. If technology were to eliminate our reliance on fossil fuels, let's consider what needs to happen next. We need:

1. A few geniuses smarter than Einstein, Hawking and Pauling.

2. Massive amounts of high risk venture capital.

3. Major technological breakthroughs.

4. No interference from the oil producing countries or the existing energy industries, lobbies, and governments.

5. Time to transition and develop the new energy infrastructure.

6. Acceptance from the public at large for whom a fundamental change is not always easy.

The "Star Trek" matter anti-matter machine may one day become reality, but for right now, we rely on nonrenewable resources that are experiencing major price hikes, as supplies dwindle.

MIND-SET #6: ENERGY IS A STATUS SYMBOL

In tropical Asia, many companies lower indoor temperatures to as low as 60 degrees Fahrenheit (15.5° C) in the summer. According to staff reporter Geoffrey Fowler, in a *Wall Street Journal* article, these companies like to send the message: "We are so luxurious, we're arctic."[15] In this region, studies indicate that 72 to 78 degrees is the optimal range for indoor comfort in the summer. With the air conditioning set to 60 degrees Fahrenheit, many employees are forced to wear coats and use space heaters to stay warm in the hot summer months. Furthermore, Fowler goes on to point out that: "Cinemas and restaurants even rent out shawls to customers rather than turn down the air conditioning when it becomes fashionably, but uncomfortably cold."[16] These are good examples of a societal precondition that more is better; the symbolic consumption of electricity is in this case, a status symbol.

Through education, more people can become aware of the consequences of wasting energy. Needlessly using energy is not benign; it wastes money and pollutes the environment. Lest there be any doubt, intentionally over cooling a building in the summer and then using space heaters to raise the temperature to a comfortable temperature is wasteful. Fortunately, status symbols are not always static and one-dimensional; they can be fluid, as we will learn in later chapters. A wise energy policy can make conservation the preferred status symbol. A wide range of people, from the low profile millionaire next door to highly public figures, turn off the lights and do not consider it a failure, but rather, the right thing to do for themselves and the environment. Running heating and cooling systems simultaneously, can, in the future become an embarrassment, not a status symbol.

MIND-SET # 7: CHANGE CAN BE HARD

Attitude is everything and, when it comes to energy savings, a cultural change is needed. We must keep our minds open

to these new ideas and learn to modify our behavior. Studies have shown that a conscientious effort by individuals to avoid needless waste of energy and water can yield savings of 10 percent.

—Vice-Admiral Garnett
Vice-Chief of Defence Staff
Canadian Department of National Defence

Consider that during the Arab Oil Embargo, which commenced in 1973, people started switching to more fuel-efficient cars. As gasoline prices stabilized, many people returned to larger cars and even sport utility vehicles. So this raises the question, did we change, or simply adapt to a temporary situation? Individual answers will vary, but we learned we should never underestimate the effort required to change energy-consuming behavior.

One way to make change easier is to minimize it and only change what is necessary. In stable organizations, core values can stay fixed, while operating practices, strategies, and processes change in response to changing realities. Ever increasing energy costs are a reality that will necessitate change, but we may not need to change the organizational culture to reduce energy costs. Energy saving initiatives can often be adapted to the organizational culture. Changing the culture may be enchanting and self-gratifying, but if we only need to change a few processes, enhance communications, and track results to get the job done, then let's just do that.

If cultural change is necessary, don't shy away from it either. If the energy awareness campaign is persistent, if posters and reminders to save energy are fresh and changed often, the results will be positive. We will have the benefits of both short-term change and long-term adaptation. Safety training is a good analogy. A safety program does not just tell a worker to wear a hard hat one time, but rather continually reinforces and monitors behavior. Energy awareness requires the same persistence; one meeting seldom accomplishes change.

MIND-SET #8: SAVING ENERGY IS INCONVENIENT

Many people have the mind-set that conserving energy is inconvenient or requires a tremendous sacrifice. In truth, it can be inconvenient, though it doesn't necessarily need to be. Consider the following example:

> ***In another plant, just labeling the light-switches, so everyone could see which switches controlled which lights, saved $30,000 in the first year."***[17]

At a minimum, the organization should "take the rocks off the rails" and facilitate individual efforts. Organizations can make it much easier for employees to save energy by correctly labeling switches, providing employee-suggestion boxes, fostering communications, and rewarding the right behavior.

At work, employees often just don't know how to save energy or were never asked for input. No one ever explained the ramifications of wasting energy, such as cost overruns or needless pollution. Organizations that make it easy to save energy, reinforce the right behavior, and solicit employee involvement experience significant energy savings. Take the time to ask what you can do to save energy, or the waste continues.

MIND-SET #9: IT'S IMPOSSIBLE TO CONSERVE NATURAL RESOURCES

Bounce Back Effect, Rebound Effect, Takeback Effect, and *Offsetting Behavior* are terms that describe a condition where improved efficiency does not generate an equal financial savings. It is possible that as conservation methods reduce the demand for a given commodity, such as oil, there will be less demand for the commodity and its sales will drop. As sales of the commodity shrink, its corresponding price will also drop, making it more affordable and available to larger markets, hence increasing demand for the commodity. In practice, this means if the United States used more fuel-efficient cars and consequently purchased less oil, the citizens of an emerging market such as China will utilize the excess oil at a cheaper price. Alternatively, should European governments impose large gasoline taxes, Europeans will buy smaller cars and drive less, but Americans will take advantage of lower gas prices to drive large sport utility vehicles. There is some evidence this may be true,[18] but in this case, it is a macro issue. The energy your company saves improves the organization's bottom line and reduces local pollution. To be absolutely clear, conservation will benefit you as an individual and an organization. The fact that other companies may consume the energy we saved, while it is abundant, is a separate issue.

It is also well documented that the first oil embargo in the early 1970s sparked a conservation trend that did reduce energy intensity. In the United States, between 1945 and 1973, consumption of petroleum products rose at an average annual rate of 4-1/2 percent. After the oil embargo, between 1973 and 2004, oil consumption slowed, on average, to only 1/2 percent per year, far short of the rise in real GDP.[19] Alan Greenspan noted in a speech given in 2005 that "...part of the decline in this ratio is due to improved energy conservation for a given set of economic activities, including greater home insulation, better gasoline mileage, more efficient machinery, and streamlined production processes. These trends have been ongoing but have likely intensified of late with the sharp, recent increases in oil prices."[20] Other countries experienced a similar effect. For example, in France oil use was still 10% lower in 2005 than it was three decades earlier, and its energy intensity in 2005 was 30% lower than in 1973.[21]

Of course, we do need to watch for some micro economic consequences. Documentation from residential studies indicates that the savings of conservation devices, such as low water showers and low flush toilets, can be offset by using the saved water to wash the car, or perhaps to water the lawn a bit longer. Residential occupants have installed high efficiency lighting, then let the lights burn longer (as it is now cheap to do so), and thus eroded their savings. The theory is that the occupant knows he is saving substantial money on utility bills, and hence, uses more energy on other applications. Fortunately, energy awareness programs are not as susceptible to this behavioral phenomenon. The goal of the awareness program is to continuously reduce energy consumption, not just deploy energy saving technology.

MIND-SET #10: WHAT'S IN IT FOR ME?

This may sound a bit selfish, or mercenary, but often the users of electricity will ask, "What's in it for me, if I conserve energy?" Often, due to an organizational pitfall called the spilt incentive, there may be no direct economic rewards for saving energy. Students that turn off classroom lights pay the same tuition as students that don't turn off lights. Workers that put computers in low power modes during lunch make the same as workers that don't. Since action, decisions, and change require a greater justification than inaction or maintaining the status quo; it is important to answer the question, "what is in it for me?" At a minimum, the answers are, a better environment, a more competitive organization, a more reliable energy supply, and potentially lower energy prices at home.

Of course, rewards do sweeten the pot and can increase energy savings. The rewards do not need to be financial; they can simply be recognition and reinforcement of energy conservation activities. One of the best organizations at this is the US Department of Energy Federal Energy Management Program (FEMP). They recognize and publish the pictures of Energy Award winners. This recognition, in turn, motivates others to save energy. In another instance, some universities share a small portion of the savings with participants and allow the students to use the money however they see fit. In a business setting, this reward system can be as simple as an Energy Champion Coffee Cup. If you feel rewards are necessary, be sure to utilize them as necessary.

MIND-SET #11: MY ENERGY BILL IS TOO SMALL TO WORRY ABOUT.

Companies tend to focus on the largest expenses. The big-ticket items vary by industry and can include employee salaries, land and buildings, machinery, raw materials, sales and marketing expenses, or even research and development budgets. When compared to other expenses, it is easy for individuals and companies to consider energy a small and fixed cost expense.

There are a few easy things we can do to shift this mind-set. We can consider our energy bill, or at least a portion of our energy bill, a controllable cost. Energy is a controllable operating cost. For example, at Simon Property Group, energy represents about six percent of their total consolidated revenue, a relatively small number. However, viewed differently, their energy cost also represents 30% of their controllable operating costs, a relatively large number. With energy surfacing at 30% of controllable costs, it is difficult to ignore energy costs.

Another common tactic is to consider the impact of energy saving on corporate profitability or other simple metric. Some good examples come from U.S. Environmental Protection Agency, which calculated that reducing energy by ten percent in a hotel has the same financial benefit as selling about 930 additional room nights a year. Or, for every 10 percent reduction in energy costs, a supermarket can boost profit margins by close to 6 percent. At another company, where energy was only a few percent of total business costs, basic energy savings were calculated to increase net earning by over 55%. This made energy saving a much higher priority. Viewing energy costs in a relevant light helps motivate employee involvement.

SUMMARY

Our attitudes and beliefs can influence our energy consumption. It is important to understand and address the human aspects of why some of us are more apt to conserve energy. This understanding will help us understand our own energy consuming behavior and help us effectively motivate others to reduce energy consumption.

Our motivation to save energy tends to decrease with:

- Coercion to save energy.
- Over-zealous or fringe conservation goals.
- A feeling that the problem is too large for us to impact or change.
- A belief that technology will save the day.
- An increase in the hassle factor or difficulty in saving energy.
- Belief that energy is a small and fixed cost.

Our motivation to save energy tends to increase with:

- Training and information on the financial and environmental consequences of energy consumption.
- Team work and peer pressure.
- Measurement and recognition of personal contributions towards energy efficiency improvements.
- Belief that the organization is genuinely interested in reducing energy costs.
- Incentives or rewards for exceptional efforts, and consequences for failure.

References

1. Jeremy Cherfas, "Skeptics and Visionaries Examine Energy Saving," *Science*, January 1991, Vol 251, P. 251.
2. Chapter 6 discussed this in more detail.
3. Loren Lutzenhiser, "Social and Behavioral Aspects of Energy Use," *Annual Review of Energy and the Environment*, Vol 18, 1993, P. 252.
4. International Energy Agency, "SAVING ELECTRICITY IN A HURRY," 2005 http://www.iea.org/Textbase/publications/free_new_Desc.asp?PUBS_ID=1554 (June 22, 2006)
5. ibid
6. ibid
7. ibid
8. http://www.sustainability-indexes.com, (June 22, 2006)
9. http://www.corporateknights.ca/reports/, (June 22, 2006)
10. Innovest Value Investors, "Energy Management & Investor Returns: The Real

Estate Sector," October 2002, http://www.energystar.gov/index.cfm?c=business.bus_good_business (June 22, 2006)

11. John Douglas, "The Energy Efficient Office," Electric Power Research Institute *EPRI Journal*, July/August 1994, Page 18 and "Bright Ideas For Saving Energy At American Express While Protecting Our Environment," A brochure by American Express and Con Edison, October 1991.
12. National Environmental Education & Training Foundation (NEETF), "Americas' 'Low Energy IQ:' A Risk to Our Energy Future: Why America Needs a Refresher Course on Energy," The Tenth Annual National Report Card: Energy Knowledge, Attitudes, and Behavior, August 2002, P 16. NEETF was Chartered by Congress in 1990, and is a private, non-profit organization.
13. Ross Petras and Kathryn Petras, *The Lexicon of Stupidity*, (New York, Workman Publishing, 2005), P. 233
14. Petras and Petras, P. 231
15. Geoffrey A. Fowler, "Kind of Blue: In Asia, Elite Offices Show Off With Icy Temperatures," *The Wall Street Journal*, August 25, 2005.
16. ibid
17. Amory B. Lovins and L. Hunter Lovins, "Climate: Making Sense and Making Money," November 13, 1997, P. 14.
18. Herbert Inhaber, *Why Energy Conservation Fails*, (Westport, CT Quorum Books, 1997)
19. Chairman Alan Greenspan before the Japan Business Federation, the Japan Chamber of Commerce and Industry, and the Japan Association of Corporate Executives, Tokyo, Japan, October 17, 2005
20. ibid
21. EUROPEAN COMMISSION, "Doing more with less, Green Paper on energy efficiency," June 2005, http://ec.europa.eu/energy/efficiency/index_en.htm, (June 22, 2006)

Chapter 3

Common Electrical Myths That Increase Energy Costs

> ***Taking a few elementary steps towards greater efficiency can save businesses millions of dollars in a short time. Most energy efficiency measures have a relatively quick payback, and many cost nothing to implement. BT, the UK telecommunications group saved £119 ($214m) between 1991 and 2004, and by more efficient use of its transport, saved £421m in the same period.***
>
> —Fiona Harvey
> *Financial Times*

When meeting people for the first time, a classic conversation starter is the simple question, "What do you do?" Sometimes, I answer, "I study why people don't turn off lights and other equipment." From this point on, people take the opportunity to ask me energy related questions. A doctor once asked, "my service company says it's cheaper to run the air conditioners constantly throughout the day, is this true?" One lawyer proclaimed she always turned off the lights in her peer's offices. However, her fellow lawyers lacked her enthusiasm of energy conservation. She then proceeded to inquire if it was true that turning off lights used more energy than leaving them on? Often confronted by her argumentative coworkers she was unsure as to the validity of her actions.

I've found that highly educated professionals and die-hard conservationists pass up opportunities to save energy as the result of misinformation. Some academics call this imperfect knowledge. So let's examine nine prevalent myths and put them in the right energy efficient light.

MYTH #1: THE INRUSH MYTH:

Myth or Fact? It is better to leave lights on, since they use more energy when turned back on.

Answer: The lights in our offices, which are typically incandescent or fluorescent lights, can use as much as five times more the energy when initially turned on. However, this higher energy use lasts for only a fraction of a second, maybe 0.05 seconds, so it does not substantially increase costs. Accounting for the start-up current, typically called inrush current, a fluorescent bulb still uses 500 times more energy if left on for 15 minutes than if turned off. The high inrush current should not prevent turning off the lights.

MYTH #2: THE LIFE CYCLE MYTH:

Myth or Fact? It is better to leave the lights on, since turning the lights on and off shortens the bulb's life and will cause it to burn out faster.

Answer: It is true that each time you switch fluorescent lights on the life expectancy of the bulb, especially with older types of bulbs, decreases slightly. However, it is also true that each minute the bulb is off adds to the life expectancy of the bulb. The general rule of thumb is to turn fluorescent lights off, if they will remain off for approximately two to fifteen minutes. The general rule of thumb for incandescent lights (the standard light bulb found in most homes) is to turn them off as often as possible.

High Intensity Discharge Lamps (HID) are another alternative to fluorescents and incandescent lamps. HID lighting systems are used in a number of facilities, including offices, streets, stadiums, factories, schools, stores, airports, and shopping malls. There are four types of HIDs: high and low pressure sodium, metal halide, and mercury vapor. HID lamps need several minutes to cool off and restart. Therefore, most HID lamps should remain off for longer periods of time than fluorescent lights. It depends on the size and type of HID bulb, but off-time recommendations are generally one hour, but occasionally as few as 15 minutes.

The cycling of lights should be coordinated with your facilities or operations department. A qualified engineer can vet the general rules of thumb for your site. The engineer can analyze your cost of energy, the price of new bulbs, and the labor to replace them. In some cases, this may be a worthy endeavor. Some facilities, where energy is particularly expensive, now recommend turning off both fluorescent and incandescent lamps regardless of the time duration. The last one out should turn off the lights.

MYTH #3: THE AIR CONDITIONER MYTH

Myth or Fact? Many people believe that companies should not turn off the heating or air conditioning after working hours or on weekends as it takes more energy to return the building to the proper temperature.

Answer: Simply not true. The heating or air conditioning unit will have less total run time hours when shut completely off or when the thermostat is set to a less comfortable level when the space is unoccupied. The fewer hours it runs, the less energy you use. You can also save energy by setting back your home units while at work. Keep in mind though, that the temperature in large spaces tends to change slowly. It is often necessary to pre-heat or pre-cool the space before the arrival of personnel. The pre-heating or cooling time can vary depending on the difference between the inside and outside temperatures, and many other factors. A common mistake is to keep the pre-conditioning time constant and longer than necessary, such as always starting the office air conditioner Sunday night at midnight.

MYTH #4: THE SCREENSAVERS MYTH

Myth or Fact? Computer screensavers save energy.

Answer: Simply not true. If the computer screen, commonly called a monitor, presents the same image in the same place for a long period of time the monitor can develop a "memory" of the image and continue to display a dim version of the image. This condition is called "burn-in" and is permanent. You might be familiar with the effect of burn-in at the automatic teller machines at your local bank, the bank logo or other text may appear as a faint image at all times. A screen saver provides a constantly changing image on the monitor when the computer has been idle for a predetermined amount of time. A constantly changing image prevents burn-in.

Many new monitors are better at preventing burn-in and do not require a screen saver. More to the point, while screensavers prevent burn-in, they do not save energy. Even setting a screen saver to "blank screen" in lieu of a moving display does not save much monitor energy. The monitor still uses almost full power. Complicated screen savers require a lot of computing resources and can actually increase the amount of heat the computer generates and the energy it requires. In addition, the screen savers changing image can keep the monitor from going into a sleep or energy saving mode and automatically dimming.

Ideally, we should turn the monitor off when not in use, or enable the computer energy management features. Consult you computer department for help.

MYTH #5: THE ENERGY EFFICIENCY MYTH

Myth or Fact? Advances in energy efficiency means we do not need to turn equipment off.

Answer: The U.S. Department of Energy estimates that office equipment will be the fastest growing commercial electrical energy end use between 1998 and 2020. Turning devices off is more important than ever. It's true that some newer devices, such as lighting and flat panel computer screens, do require less energy, but energy intensity, that is the amount of energy we use per square foot of office space continues to climb rapidly. According to 2003 projections by the Department of Energy, annual energy use by personal computers is expected to grow 3% per year, and energy use among other types of office equipment is expected to grow 4.2% annually; this growth is in spite of improvements in energy efficiency.[1]

The gains in efficiency of some devices are offset by the increased energy consumption of other devices and the overall increase in the total number of energy using devices. Therefore, it is more important than ever to turn off unneeded devices.

MYTH #6: THE HEAT EXCHANGE MYTH

Myth or Fact? I don't need to worry about leaving on lights in the winter as they help heat the building and therefore reduce heating costs.

Answer: Usually false, but it depends. It's true that most of the energy going to lights, computers, or process equipment ends up as heat.[2] Our bodies also produce a lot of heat. Each of us produces about 500 to 600 Btu of heat per hour. Because of this internal heat generation (and the heat from sunlight hitting the building), many building owners are constantly cooling their buildings, even in the winter. This is an important point that bears repeating: some large office buildings in cold climates need to cool their buildings even when it is cold outside.

Therefore, often this myth is not true; running lights and equipment in the winter can still increase the total energy costs. In addition, when the indoor air temperature is too cold, it is usually more economical to heat the building with the central heating system than leaving on lights and equipment. If in doubt, check with your facilities department to determine the heating and cooling characteristics of your building.

MYTH #7: THE PHANTOM LOAD MYTH

Myth or Fact? When electronic devices are off, they are not using energy.

Answer:

> *Many idle electronics—TVs, VCRs, DVD and CD players, cordless phones, microwaves—use energy even when switched off to keep display clocks lit and memory chips and remote controls working. Nationally, these energy "vampires" use 5 percent of our domestic energy and cost consumers more than $3 billion annually.*
>
> —Lawrence Berkeley National Laboratory

An astonishing number of electrical devices consume power when switched off. Worldwide, this standby power is estimated to account for as much as 1% of global green house emissions.[3] A study by Alan Meier of Lawrence Berkeley National Laboratory found that, based on rough assumptions, office equipment in low-power modes may be responsible for 1.1 billion kWh a year of energy consumption just in California.[4] In stand-by mode an average computer may use 15 watts, a halogen desk lamp may use 3 watts, and a small copy machine between 45 to 60 watts. This adds up. There is some reason to believe that the energy use of equipment in low power modes is the fastest growing component of energy use.

Often ignored are the power supplies (sometimes called "wall warts") that enable equipment to run on a higher plug voltage. These power supplies consume power even when the associated equipment is off or disconnected. For example, my laptop computer cord has a built in power supply. When I travel I take my laptop with me, but leave the power cord at home and plugged in, the cord continues to run up my home electrical bill. Power supplies like this can consume 1 to 3 watts when the computer is off. Often these devices feel warm to the touch. A good rule of thumb is the hotter the device is, the more energy it consumes. Each watt consumed in this stand-by mode can total to 8.74 kWh per year, and costs about one dollar a year. My laptop computer cord may cost me $3.00 a year in direct home energy charges even when the laptop is off. Add in my phone charger, television, microwave, fax machine, printer, other similar devices, and I pay over $100 a year powering off devices. A conservative estimate of residential standby power in the United States is about 70 watts per home, which is about 600 kWh a year or 10% of total residential use.[5]

There is an effort to reduce the standby power consumption. ENERGY STAR is a government-backed program helping protect the environment through better energy efficiency. Created by the U.S. Environmental Protection Agency in 1992, it now sets voluntary international standards for energy efficient equipment. ENERGY STAR is working with manufactures to reduce standby power.

Market forces can also drive manufactures to reduce standby power. For example, President Bush released an Executive Order in July 2001 in which he stated: "Each agency, when it purchases commercially available, off-the-shelf products that use external standby power devices, or that contain an internal standby power function, shall purchase products that use no more than one watt in their standby power consuming mode. If such products are not available, agencies shall purchase products with the lowest standby power wattage while in their standby power consuming mode. Agencies shall adhere to these requirements, when life-cycle cost-effective and practicable and where the relevant product's utility and performance are not compromised as a result." As the largest energy consumer in the United States, the federal government is leading by example. To assist in the effort, Lawrence Berkeley National Laboratory currently offers a standby electricity website listing products with low standby power, including office equipment, consumer electronics, and appliances.

MYTH #8: THE AUTO SHUTDOWN MYTH

Myth or Fact? Frequently shutting off computers and monitors will cause premature failure.

Answer: Contrary to popular opinion, orderly shutdowns do not damage computers. Modern computers can handle 40,000 on-off cycles before failure.[6] Shutting down computers that are idle for 20 minutes is beneficial to the energy bill. However, in practice, this is often not realistic. Computers can take a lengthy time to restart. Decreased productivity can quickly cancel out energy savings. Fortunately, there is a painless alternative.

Power management is a process that allows computers and monitors to enter low-power states when sitting idle for a predetermined amount of time. Simply hitting the keyboard or moving the mouse will wake the computer or monitor from its low-power "sleep mode." It is a good idea to utilize the power management features on your computer to power down computers left inactive for 15 minutes.

The U.S. Environmental Protection Agency indicates that computer power management can save $15 to $45 per monitor annually and monitor power management can save $10 to $30 per monitor per year. Added together, a desktop computer with monitor can save $25 to $75 annually with the power management features enabled.[7] Of course, this amount varies by the type of computer and your local energy costs. In the U.S., ENERGY STAR estimates that 45 percent of computer monitors do not take advantage of built-in sleep features.[8] One EPA estimate is that businesses and organizations across the nation can save a total of more than $900 million and 11 billion kilowatt-hours with monitor power management.[9] In total, the energy wasted by computers and monitors costs U.S. organizations about $1.5 billion every year.[10] The EPA also states that computers and monitors use more electrical energy than all other forms of office equipment combined. For more information, visit the EPA power management website.

Of course, "sleeping" and "off" are two different things. A low power state, or standby power, still uses more energy than a device that is off. And "off" still uses more power than unplugged. Notice in Figure 3.1 that unplugged may be the only power state that uses no energy.[11] Connecting your computer and monitor to a power strip and de-energizing the power strip at night will save energy, but check with your technical support people first. In addition to the power supply consuming power, another reason computers draw energy even when turned off is that the motherboard always has some control power. This allows the computer to awaken and turn on through a network or modem connection. Organizations will sometime use this feature to upgrade software and perform maintenance on computers during the night.

Often overlooked, computer accessories such as speakers connected to the computer continue to draw power when the computer is sleeping or turned off. These standby loads typically vary from 1 to 8 watts. Computer power management features will not effectively reduce these loads. If possible, attach these electronic devices to a plug-in power strip with a switch. Turn off the power strip at the end of the day. There is little down side to doing this. A power strip with six outlets and a switch costs only a few dollars. As electrical energy costs rise, and electronic devices proliferate, the amount we spend on standby power grows steadily

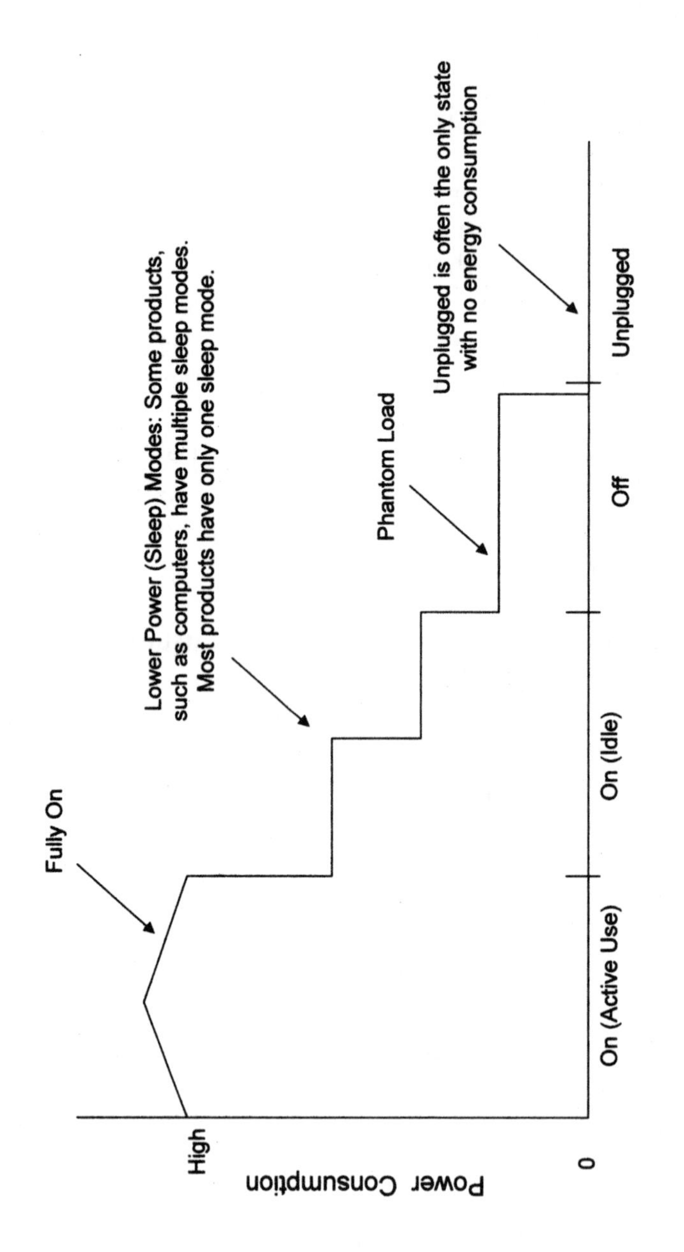

Figure 3.1: Relative energy consumption in various power modes.

MYTH #9: ENERGY AWARENESS EFFORTS ARE SHORT LIVED

Myth or Fact? Our efforts to turn off devices may save lots of money in the short run, but we will eventually lose interest in energy conservation and revert to our prior consumption patterns.

Answer: Energy awareness programs, if not properly executed, can and will suffer from reduced effectiveness over time. One study showed that placing stickers on computers and light switches reduced energy costs by a total of 14% and 15% respectively over an eight-week period, but the savings decreased and eventually died out over the same time period.[12] Another study indicated that just replacing existing lights with energy efficient bulbs and the attention the project created caused occupant behavior changes that saved energy for at least 11 months, though savings gradually decreased the entire time. The study terminated at 11 months, so it's not clear how long the savings persisted.[13]

Simply placing stickers on switches is not a comprehensive energy awareness program, nor is the act of observing lighting retrofits. Static stickers and posters alone are only effective if the desired outcome is short term. If a goal is to reduce energy use for a small period, say a couple of months, to deal with a temporary crisis, or a summer peak energy demand period, then buying posters or stickers will be cost effective as they are relatively inexpensive compared to the electric bill. Using stickers that can be easily removed and replaced with a different sticker in another crisis is desirable as this keeps the messaging fresh.

If the goal is long-term sustainable savings or increased savings year over year, a more comprehensive approach is needed. Continuous improvement requires a continuous process.

Energy saving is a hard grind but absolutely essential. Proper housekeeping is vital and there is a constant need to impress on people that they have to turn off taps, switch off lights and switch off machines and services. There is an ongoing task to double check.

—John O'Leary, Energy Manager, Waterford Crystal[14]

We will see in later chapters that leaders can create energy saving changes that are sustained over time. Luckily, there is a proven methodology to maintain the savings indefinitely. It's a challenge, but it can be done.

SUMMARY

Myths and misconceptions increase energy consumption. Whenever possible:

- Turn off unneeded lights.
- Accept that turning off office equipment is more important than ever. Office equipment is the fastest growing commercial segment of energy end use.
- Understand that each light, computer, or energy-consuming device gives off heat. This heat must be removed from buildings through energy intensive cooling systems, often even on the cold winter days.
- Take advantage of energy saving "sleep" or power down modes on computers and copy machines. The "sleep" modes save some power, but not as much as turning the device off completely.
- Recognize that many devices still consume power even when switched off. Consider unplugging the devices, or connecting devices to a power strip and turning off the power strip on nights and weekends.
- Energy awareness will consistently drive cost savings, if we stay vigilant.

References

1. Judy A. Roberson, Carrie A. Webber, Marla C. McWhinney, Richard E. Brown, Margaret J. Pinckard, John F. Busch, "After-hours Power Status of Office Equipment and Inventory of Miscellaneous Plug-Load Equipment," January 2004, http://enduse.lbl.gov/Info/53729-abstract.html, (June 22, 2006)

2. Dr. Wayne C Turner, "Common Sense Approaches," *Energy Engineering*, Vol. 102, No 2, 2005, P. 5.
3. Energy Star, "The Quality of our Environment is Everyone's Responsibility," June 7, 2005, http://www.energystar.gov.au/about/index.html, (June 22, 2006)
4. Alan Meier, "RESEARCH RECOMMENDATIONS TO ACHIEVE ENERGY SAVINGS FOR ELECTRONIC EQUIPMENT OPERATING IN LOW POWER MODES; A Summary of Previous Project Work and Identification of Future Opportunities," LBNL-51546, prepared for Donald Auman, California Energy Commission, September 30, 2000 and Alan Meier of Lawrence Berkley National Laboratory, e-mail to author, March 12, 2006. For more information see; Benoit Lebot, Alan Meier, and Alain Anglade, "Global Implications of Standby Power Use," To be published in *The Proceeding of ACEEE Summer Study on Energy Efficiency in Buildings.* Also published as LBNL-46019, June 2000
5. ibid
6. Rocky Mountain Institute, "Home Energy Brief #7 Computers and Peripherals" http://www.rmi.org/images/other/Energy/E04-17_HEB7Electronics.pdf, (June 22, 2006)
7. EPA ENERGY STAR Power Management website, http://www.energystar.gov/index.cfm?c=power_mgt.pr_power_management. (June 22, 2006)
8. Citigroup Inc. press release, "U.S. EPA Recognizes Citigroup for protecting the Environment through Participation in Energy Star® 2003 Million Monitor Drive"
9. ENERGY STAR press release "More Sleep Will Increase Cisco's Bottom Line by $1 Million a Year."
10. ENERGY STAR news room, "Sleep is Good: For Computer Monitors and your Bottom Line."
11. ibid
12. Guy R. Newsham and Dale K. Tiller, "The Energy Consumption of desktop Computers: Measurement and savings Potential," *IEEE Transactions on Industry Applications*, Vol. 30, No. 4, July/August 1994.
13. W.D. Chvala, Jr., R.R. Wahlstrom, M.A. Halverson, "Persistence of Energy Savings of Lighting Retrofit Technologies at the Forrestal Building," Prepared for the U.S. Department of Energy, Federal Energy Management Program under contract DE-AC06-76RLO 1830. April, 1995.
14. From a booklet by the UK Department of the Environment, Transportation and the Regions, "" March 1999, P.26.

Chapter 4

Organizational Pitfalls that Increase Energy Consumption

> *A typical soda machine uses 120 watts of non-essential lighting, costing more than $60 per machine each year. If only one tenth of the 400,000 Department of Defense buildings took out a soda machine lamp, over $2.4 million could be saved annually!*
>
> —U.S. Navy

Organizations utilize established processes and procedures to deal with day-to-day activities. Unfortunately, these processes can have unintended consequences and increase energy consumption. We will call these situations "organizational pitfalls." Let us consider some common organizational pitfalls that impede our ability to reduce energy costs. Understanding these barriers is the first step in avoiding them. Once understood, there are a half-dozen obvious fixes to hurdle these obstacles.

PITFALL #1: SPLIT INCENTIVE

One barrier to energy efficiency projects is the "split incentive." This occurs when the party benefiting from the energy saving efforts is not the party implementing or financing the initiative. Confused? Consider the "soda machine" quotation at the beginning of this chapter. Typically, the company that buys, owns, and operates vending machines is a third party company that specializes in vending operations. However, the facility where the machine is "plugged-in" pays the electric bill. When buying a soda machine, why would the vending company pay more for an energy efficient machine when they do not pay the electric bills? They would never recover the extra investment. The energy savings all go straight to the facility that pays the electric bills. In this example, energy efficiency is not a rational business decision for the vending company. This may seem like a trivial example, but one estimate puts annual electricity consumption of refrigerated vending machines at about 7.5 billion kilowatt-hours per year.

The spilt incentive barrier is especially prevalent in office and retail buildings. In the United States, more than half of the office space is rented. If the utility charge is included in the rent and the tenant does not receive a utility bill, the tenant has no economic grounds to pay more for highly energy efficient office equipment or even turn off the lights, as the property owner pays the utilities and receives the savings. Conversely, if the tenant pays the utility costs, the property owner has no economic incentive to install a highly efficient air conditioning or lighting system, as the property owner needs to pay more for the energy efficient appliances, while the tenant benefits from the lower bills. This split incentive results in higher energy bills for one or the other.

Split incentive barrier represents a legitimate obstacle to even small energy initiatives, such as turning off the lights. Why turn off the light if some other company is paying the bill? There are six common solutions to this pitfall.

The Fix

1. Tenants and property owners simply work together to align their interest and construct a win-win arrangement; this may involve modifying the lease.

2. Performance-based contracts are an effective solution. There are many forms of performance-based contracts; the common theme

is that an outside company pays the full cost of energy efficiency improvement. Once the equipment is installed, smaller energy bills are received from the utility, and an amount equivalent to that monthly energy savings is paid to the outside company until the project costs and fees are recovered. Once project costs are recovered, the person paying the energy bill has a much lower energy bill, and the building owner has a building with newer energy efficient equipment and satisfied tenants.

3. Consider starting with no-cost or low-cost opportunities such as an energy awareness program. Energy awareness programs financed by the building owner, when properly done, increase tenant satisfaction and improve the corporate image of the building owner. A few quality posters in lobbies, vending, and other common areas can advertise the property owner's commitment to energy efficiency. The goal is for energy efficiency rewards to go to the correct party.

4. The government can raise the minimum efficiency standards for equipment. This forces the purchase of energy efficient equipment, but this does nothing to drive energy saving behavior.

5. The government can mandate overall energy reduction targets. For example, the Presidential Executive Order 13123: Greening the Government Through Efficient Energy Management, mandates reduced energy use per square foot by 30 percent in 2005 and 35 percent in 2010 relative to 1985 baseline. Many states have goals to reduce energy usage in 2010 by 15% verses a year 2000 baseline. These type of mandates can drive energy saving behavior despite the spilt incentive barrier.

6. It has also been shown that focusing on reducing the harmful environmental emissions of energy consumption can induce the desired energy behavior, despite the financial aspects of the spilt incentive barrier. We'll review this in more detail in Chapter 7.

PITFALL #2: MASTER METERING

This pitfall is similar to four friends going to dinner. Three consume the $9.99 dinner special, and one gets the $70 steak and lobster platter, dessert, and 64-oz Margarita special. When presented with the $160 check, the big eater says, "Let's split it four ways and all chip in $40." Likewise, many large office buildings and facilities have only one electric meter. This is known as master-metered. In this scenario, the bill is divided based on some formula. The formulas vary, but typically consist of simple square footage calculations or engineering studies of the installed equipment. Regardless of the formula, here, one tenant may conserve energy, yet not reap any benefits as their neighbor's waste energy. Master-metered complexes can consume 35% more energy than individually metered sites.[1]

Companies, business units, and individuals are not as accountable for energy usage when they are master metered and billed. When employees and businesses are not accountable for waste, there will be lots of waste.

The Fix

The alternative to master metering is sub-metering. In sub-metering, each tenant is billed individually. This holds the tenant accountable for his usage and results in energy savings. In fact, it is so effective that I have even considered installing a meter in my daughters' bedrooms.

If possible, ask to have your office, department, or cost center metered separately to track and document your energy usage. Oddly enough, many states discourage building owners from sub-meter tenants by making it illegal, or extremely difficult. New York has favorable laws for consumer installed sub-metering; California does not. In general, both the tenant and building owner want the sub-meter. One wonders, why more lawmakers do not.

PITFALL #3: NOT INVENTED HERE SYNDROME

With the "not invented here" syndrome, ideas from elsewhere are intentionally or sometimes unintentionally unwelcome. Often, people refuse to take responsibility for, or utilize something they don't completely understand. This will manifest itself as "We're different. I can see how energy awareness initiatives work for other organizations, but it just won't work for us," or "You don't understand, we already conserve as much energy as possible," or my favorite, "Energy efficiency will adversely impact product quality or customer service." I do understand these objections, even if they are false. History is full of similar business examples that ended in failure due to obstinacy of this kind:

> This 'telephone' has too many shortcomings to be seriously considered as a means of communication. The device is inherently of no value to us. —Western Union internal memo, 1876.

> While developing the technology behind the laser, American Physicist Charles Townes was approached by two Nobel-Prize-winning colleagues who told him he was wasting his time and threatening their funding. Even after the first laser was built in 1960, it was described as "a solution looking for a problem."[2]

> So we went to Atari and said, "Hey, we've got this amazing thing, even built with some of your parts, and what do you think about funding us? Or we'll give it to you. We just want to do it. Pay our salary, we'll come work for you." And they said, "No." So then we went to Hewlett-Packard, and they said, "'Hey, we don't need you. You haven't got through college yet." —Apple Computer founder Steve Jobs on attempts to get Atari and HP interested in his and Steve Wozniak's personal computer.

The Fix

If your organization has similar reactions towards an energy awareness program, rest assured there is ample evidence that energy awareness training is necessary and applicable to a diverse set of organizations.

Knowledge is the key. The more research you put into saving energy the more apparent the potential becomes. Where possible, use "not invented here" off-the-shelf training and educational tools to help keep the costs down and keep the program fresh and ongoing.

PITFALL #4: NOBODY IS FIRED FOR DESIGNING ENERGY INEFFICIENT SYSTEMS

When designing manufacturing processes or buildings systems, engineers tend to oversize motors and other pieces of equipment. If the engineer specified equipment that was too small, the process or building would suffer noticeable results. When equipment is oversized, no one really knows. The drawback is, of course, higher energy bills.

The Fix

Employees should be judged, in part, on the energy efficiency of their projects. Ideally, the awareness program will make them aware of the issues, and when replacing a motor, or designing a package, they may check to see if there is a more energy efficient solution. A great example comes from Amory Lovins and L. Hunter Lovins, of the Rocky Mountain Institute, in "Climate: Making Sense and Making Money":

> *Leading American carpet maker Interface was recently building a factory in Shanghai. One of its processes required 14 pumps. The top Western specialist firm sized them to total 95 horsepower. But a fresh look by Interface/Holland's engineer Jan Schilham, applying methods learned from Singapore efficiency expert Eng Lock Lee, cut the design's pumping power to only 7 hp—a 92% or 12-fold energy saving. It also reduced the system's capital cost, and made it more compact, easier to build and maintain, and more reliable and controllable.*[3]

Engineers are good at designing systems to meet the project goals, so consider making energy efficiency a desired goal. This activity can easily carry over to other disciplines. For example,

Dannon yogurt containers had both a foil and a plastic lid, the equivalent to wearing a belt and suspenders. A simple redesign eliminating the plastic lid saved Dannon 3.6 million pounds of plastic a year. Since the primary ingredient of plastic comes from oil, Dannon's packaging change also saved a non renewable energy source.

PITFALL #5: ZERO COMPLAINTS POLICY— ENERGY MANAGEMENT IS NOT SEEN AS IMPORTANT

Certainly, there are no economic grounds for favoring energy efficiency over core activities, such as manufacturing or product strategy. In fact, energy may intentionally take a back seat to other activities because of conflicting goals. One common example is the zero complaint policies for building occupants. One school administrator I spoke with in Ontario said their school funding was tied to student evaluations. In this case, student comfort greatly out weighted energy conservation activities. Should students complain about anything, including the temperature or lighting levels, the schools funding would be reduced. Many office and retail companies have similar polices in effect.

The Fix

First, don't use the zero complaint policy as an excuse to waste energy. Improved tenant comfort and/or product quality can go hand-in-hand with energy efficiency. An area ripe for savings is typically the start time of the preheat or precooling cycle for buildings. Figure 4.1 shows the energy consumption pattern of a typical building. In this case, the building operators started the preheating or precooling cycle at around 1:00 A.M. This ensured that the heating ventilating and cooling system (HVAC) had time to get the building to the desired temperature prior to the tenant's arrival. After careful review and some convincing, the building operators changed the start time to 4:30 A.M. This achieved the same result, a nice air temperature at

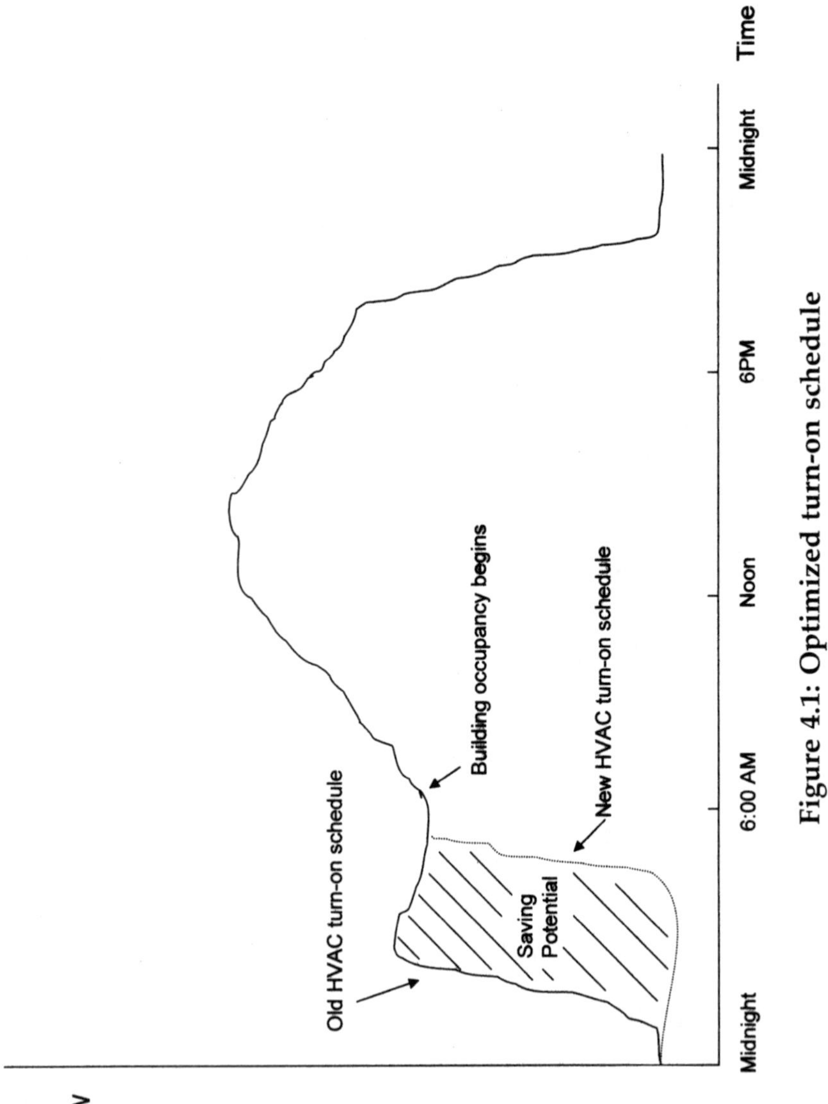

Figure 4.1: Optimized turn-on schedule

the start of the workday, but saved nearly $20,000 a year in a 300,000 square foot office building.

As energy costs continue to rise and stay sufficiently high, companies and institutions should consider energy conservation as a way to reduce expenses. A logical place to start is communicating a comprehensive energy policy, which addresses low-cost to no cost initiatives and gain the buy-in of the building occupants and equipment operators.

PITFALL #6: EXCLUSIVE RELIANCE ON TECHNOLOGY

In the debate between technological solutions and employee awareness initiatives, those arguing for technology usually assume that technological solutions have superior sustainability. As it turns out, technical approaches to energy efficiency can also seriously suffer from reduced savings over time. One study of 60 buildings found that 50% of the buildings had control problems, 40% had heating ventilation and air conditioning problems, 25% had energy management systems, economizers, and/or variable frequency drives (devices that adjust motor speeds efficiently) that were not functioning properly, and 15% had missing equipment. All of which reduced energy savings.[4]

The Fix

A comprehensive energy program that includes employee involvement. It has been shown that individuals are more willing to support technological deployments when:

- They understand the issues.
- They know the impacts associated with individual energy use.
- They know what they can do to make a difference.
- They are not inconvenienced.

Additionally, when there is an energy awareness program in place, the technology tends to perform and function optimally. Communication is the key. Even the simplest things can go wrong without proper energy communication. For example, at one site all the lights were retrofitted with energy saving bulbs. The labor to do this was expensive, but the initiative had projected a good pay back. Unfortunately, no one told the purchasing department. They continued to buy the cheapest replacement bulbs available. Over time as the energy efficient bulbs burned out, they were replaced once again with the less energy efficient bulbs wiping out the prior initiative. Clear communication to all departments involved would have prevented the problem.

PITFALL #7: THE DISEMPOWERED ENERGY MANAGER

An organization may not have an energy manager, or the energy manager may not have direct control of any funds to implement energy management initiatives. Under this last scenario, which is most typical, the energy manager must present the business case for an energy efficiency project to each facility or cost center manager and encourage them to implement the project. This lack of power, funds, authority, and management support greatly hinders the energy manager's efforts.

The Fix

If possible, link energy management to core activities. If energy management is seen to add value to core objectives, it will be harder to ignore. If local managers and employees are accountable for energy reductions, the tide turns and they seek help from the energy manager.

PITFALL #8: LITTLE INDUSTRY SUPPORT FOR ENERGY AWARENESS PROGRAMS

Traditionally, sales people play a key role in educating customers and promoting new technologies. New technologies often have high profit margins that finance the sales and marketing efforts. Energy awareness initiatives are considered low cost or no cost activities. In the no cost scenario, there is no profit motive for the energy efficiency industry to promote the energy awareness concept. Consequently, we do not have scores of sales people assisting Corporate America with the implementation of energy awareness solutions. As a result, fewer companies are aware of the opportunities. Those that are have needed to devote excessive time and resources creating their own communication collateral such as posters and stickers.

The Fix

To fill the void left by industry, governmental and nonprofit organizations have developed energy awareness tips. Communication tools and tips are available from many sources, such as the U.S. Department of Energy, Office of Energy Efficiency Natural Resources Canada, and the UK sponsored Carbon Trust. As more awareness tools become convenient and widespread, energy savings should increase. In the mean time, I have included useful links to energy awareness tools in Section II.

PITFALL #9: ENERGY MAY BE ARTIFICIALLY CHEAP AND THUS EASY TO WASTE

Some economists consider energy prices artificially low as not all the costs incurred by society are priced into the product. For example, say there is a manufacturing facility located on a river. The factory makes widgets and sells them for $5. During the manufacturing process, the factory pollutes the river, killing or rendering the fish inedible. This then will close the entire fishing industry along the river, drop the value of homes in the area, and potentially wreak havoc with the local economy. However, we, who live far away from the factory, are not affected by this environmental impact, and are usually unaware of the plight of the economy. If the market mechanisms were perfect, the price of the widget might be higher, as the local economy would be compensated for the impact of the manufacturing process. However, this rarely, if ever, happens.

As the example illustrates, some environmentalists will argue that the environmental impact of electrical generation is not included in the price we pay for products or electrical power. This artificially low energy price makes implementing many energy saving initiatives cost prohibitive, although the net benefit to society would be positive. For example, if the price of relamping an entire building with energy efficient bulbs is fixed, the pay back period for this investment decreases as the price of electricity increases. The shorter the pay back period is on an energy efficiency investment, the more likely it is to happen. As energy costs rise, organizations will have more incentive to conserve energy.

The Fix

Some countries have imposed large taxes on energy use to encourage conservation. Ideally, I would prefer to see organizations adopt responsible business practices and voluntarily reduce needless energy consumption.

PITFALL #10: CAPITAL MISALLOCATION

Organizations may understand the importance of managing energy costs, but nonetheless choose not to spend any money on energy efficiency. This may be the result of energy efficiency initiatives competing with the capital or operational budgets of other projects, or, simply the result of not considering energy projects during the budgeting cycle.

The Fix

Energy awareness initiatives do not require capital funding and are so low cost they are often considered no cost (with rounding). No cost, low cost improvements yielding 20-30% savings are a great place to start. Start with the critical success factors in chapter

8. Then use the energy awareness savings to fund initiatives suggested from your newly trained and energy conscious staff between budget cycles. Also, explore various utility and government rebates and tax credits, "guaranteed saving initiatives," or performance-based contracts offered by energy service companies.

PITFALL #11: WE ARE ALREADY ENERGY EFFICIENT

Energy managers often tell me about executives walking through facilities and touting, if not bragging, about their energy efficiency. This puts the energy manager in a tricky situation. They can spot energy waste and typically see lots of opportunities, or at least avenues to explore, yet they just don't want to disagree with the executive, or worse, be viewed as a naysayer.

The Fix

Treat energy like any other operating cost or expense. Always look for ways to reduce energy costs and keep raising the bar on internal metrics, such as energy cost per widget, or energy cost per square foot. There is a lesson in the fact that many energy efficient organizations utilize energy awareness tools to continually leverage employees and obtain further savings.

PITFALL #12: PREVIOUS ENERGY MANAGEMENT FAILURES

Commercial and industrial businesses have their share of failed energy management initiatives. The energy saving technology may have been too complicated or unreliable. Often projects may have exceeded budget or the energy saving potential was overstated. Even bad energy management experiences at home can hinder energy improvements at work. In any case, past failures often inhibit implementation of new opportunities by creating additional funding obstacles and approval processes.

The Fix

Fortunately, early innovators have proven the savings potential of energy awareness initiatives. Later in this book, we will review examples of organizations that saved hundreds of thousands of dollars, even millions of dollars, through low-cost and no-cost energy awareness initiatives. If you stick to the processes as outlined in this book, risks are minimized, and again, we are talking about extremely low cost initiatives.

PITFALL #13: INSUFFICIENT INFORMATION

Businesses often lack basic energy information. This can discourage energy management initiatives. Businesses conclude that they will need extra staff or in-house experts to analyze every possible contingency from cost over runs to production failures or tenant complaints.

The Fix

Just begin with the easiest, least expensive cost saving opportunities and build expertise. Start with low cost, no cost activities such as turning off equipment. Then, advance to a monitoring and reporting system. Generally, utility bills provide only the total energy used, couple this with the latency of the bill, and there is little useful information to reduce energy expenses. A monitoring system will remedy this situation by providing detailed and timely information. Also, consider bringing in outside consultants to analyze energy opportunities.

PITFALL #14: ORGANIZATIONAL INSTABILITY

If there is organizational turmoil such as mass layoffs, mergers, restructuring, or expected executive management changes, energy management takes a back seat.

The Fix

There are a couple of options. You can consider postponing the launch of an energy awareness campaign until the organizational stability has improved. Under certain circumstances, this may be the best approach. Employees may think the energy awareness program is disingenuous or that the program may lack support from the future management team. However, delaying the program may not always be possible. If there is a precipitating event, such as a widespread power outage, that foster appeals to curtail energy loads in a crisis, you may need to forge ahead. Alternatively, if the root cause of the organizational instability is a financial crisis, reducing energy costs may save some peoples jobs and be a motivating factor. Just be sure to account for, and potentially leverage, current events.

SUMMARY

Many organizational processes create pitfalls that can impede traditional energy management initiatives. These are:

- Split incentive.
- Master metering.
- Not invented here syndrome.
- Nobody is fired for designing energy inefficient systems.
- Zero complaints policy—energy management is not seen as important.
- Exclusive reliance on technology.
- The disempowered energy manager.
- Lack of industry support for energy awareness.
- Energy may be artificially cheap and thus easy to waste.
- Capital misallocation.
- We are already energy efficient.
- Previous energy management failures.
- Insufficient information
- Organizational instability.

With the exception of the last pitfall, organizational instability, energy awareness initiatives can provide the methodology, training, and communications techniques to help overcome these pitfalls and reduce energy costs.

References

1. Loren Lutzenhiser, "Social and Behavioral Aspects of Energy Use," *Annual Review of Energy and the Environment*, Vol. 18, 1993, P. 252. A residential source is Herbert E. Hirschfeld, Joseph S. Lopes, Howard Schechter, Ruth Lerner, *Residential Electrical Submetering Manual*, prepared for the New York State Energy Research and Development Authority, October 1997, revised October 2001, P 2, states "Experience shows that the change from master-metering to submetering typically reduces the consumption of electricity in apartments by 10-26%."
2. Robert Matthews, "The atom bombshell that is splitting opinion," *Financial Times*, March 10, 2006, P. 7.
3. Amory B. Lovins and L. Hunter Lovins, "Climate: Making Sense and Making Money," November 13, 1997, P 5.
4. Floyd E. Barwig, Iowa Energy center, John M. House, Iowa Energy center, Curtis J. Klaasen, Iowa Energy Center, Morteza M. Ardehali, KN Toosi University of Technology, Theodore F. Smith, The University of Iowa, "The National Building Controls Information Program," *Proceedings from the 2002 ACEEE Summer Study on Energy-Efficiency in Building*, American Council for an Energy-Efficient Economy, Washington, D.C., Vol. 3, 2002, P 1-14.

Chapter 5

How Employee Behavior Affects Other Common Energy Management Initiatives

> ***By changing employees attitudes and behavior, it is possible to significantly reduce energy use and contribute to the savings achieved through technical measures.[1]***
>
> —Office of Energy Efficiency, Natural Resources Canada

To maximize energy savings in existing facilities, it is best to adopt a comprehensive energy plan that incorporates four initiatives: energy saving technology, proper maintenance, metering, and energy awareness. Mastery of the first three initiatives is well beyond the scope of this book. Nevertheless, it is worth discussing how our behavior affects these other initiatives.

COMPREHENSIVE COMPONENT #1: TECHNOLOGY DEPLOYMENT

Lighting controls, variable frequency drives, environmental control systems, and numerous other technologies have greatly reduced energy consumption. Unfortunately, technology alone cannot solve all our energy problems. Technology is not yet cost effective at eliminating all types of needless energy consumption.

Where technology is deployed, engineers have compiled a substantial and compelling body of evidence that the behavior of employees can have a profound affect on energy savings derived from new technologies. That's right, our behavior, considerably more often than you might expect, can have a disastrous influence on technology. People sometimes, if not usually, bypass or otherwise alter the design intent. One energy manager told me automatic time clocks were typically disabled within days. Have you ever wondered why the forecasted return on investment (ROI) of many energy saving projects did not live up to expectations? Lack of employee training, motivation, or buy-in can be the cause.

A literature review of the relationship between control-related problems and energy consumption was published in a paper, "The National Buildings Control Information Program," by Floyd E. Barwig et al, of the Iowa Energy Center.[2] As shown in Figure 5.1 they found:

- **29% of the problems were human factor related,**
- 32% of the problems were software related,
- 26% of the problems were hardware related, and
- 13% of the problems were unspecified.

Barwig, et al, further delineated the motivation behind the human factor issues into four categories as follows: operator error, operator interference, operator unawareness, and operator indifference and ranked the occurrence rate as shown in Figure 5.2[3]:

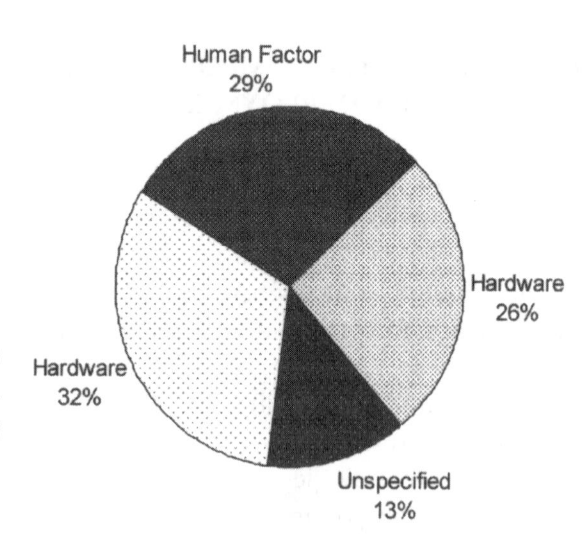

Figure 5.1: Cause of control-related problems affecting energy consumption.

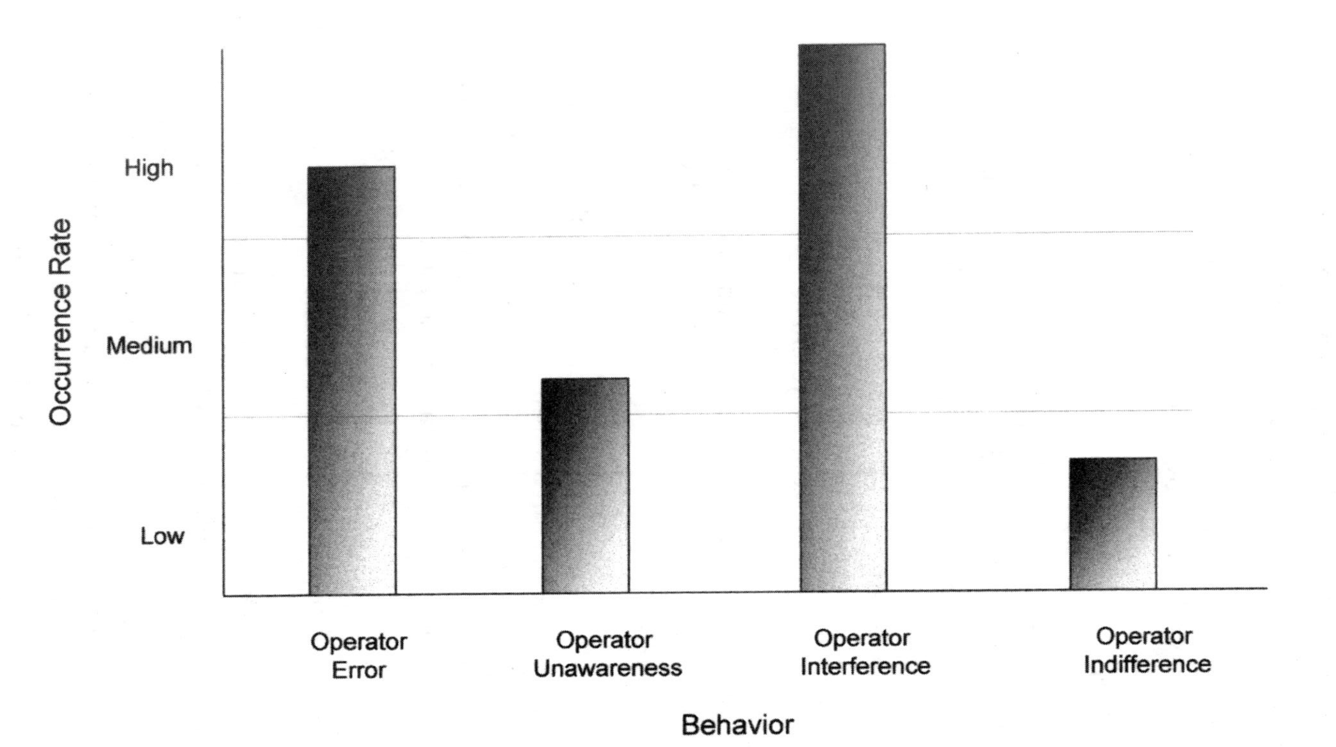

Figure 5.2: Motivation behind human factor issues

Barwig used the following descriptions for the delineation in Figure 5.2:

- Operator Error—Refers to the unintentional changes to the control system made by the operator during routine operation and maintenance that result in improper operation of a system.

- Operator Unawareness—Refers to control problems arising from an operators lack of understanding or familiarity with the control system due to inadequate training.

- Operator Interference—Refers to the intentional changes to the control system made by the operator causing interference with the normal operation of the system

- Operator Indifference—Refers to any number of problems stemming from an operators apathy toward operation and maintenance.

The findings may not be exact, as causality is difficult to determine, but the point is clear: people play a major factor in maximizing savings from technology. Where technology is cost effective, organizational buy-in can greatly increase its effectiveness. The people using or operating energy intensive devices are an integral part of the improvement process. Energy awareness practices such as high quality energy training and performance tracking can often improve the return on investment of technology.[4]

Energy management is not the only field where technology is augmented with awareness campaigns. Consider again, the memorable "loose lips sink ships" campaign of the Second World War. This WWII program was an effective reminder to everyone that national security is important and an individual responsibility. So it should not come as a surprise that when faced with another wartime situation, the U.S. National Security Agency instituted another "loose lips sink ships" campaign in 2002. The new print campaign, aimed at military bases, used dramatic patriotic art of military personnel at work inscribed with slogans such as "INFORMATION SECURITY BEGINS WITH YOU." The campaign is just one part of a comprehensive security strategy; it is not designed to replace technology or equipment. We still rely on everything from satellites to supercomputers; technology plays an indispensable role. But no matter how much we spend on gadgets, human behavior continues to play a key role. The same is true with energy management; no matter how efficient the technology, the human component remains.

COMPREHENSIVE COMPONENT #2: PROPER COMMISSIONING, OPERATION AND MAINTENANCE PRACTICES

> ***For ten years, state efficiency programs have been giving us money for technology and designing technical solutions for energy efficiency. While that has been nice, much of it has missed the mark—where we really needed the help was in operations and maintenance.***
>
> —School O&M Administrator
> "Strategies for Improving Persistence of Commissioning Benefits"

Equipment must be properly installed and periodically maintained to function at its peak efficiency. Performance monitoring projects have a potential to conserve 15 to 30% of energy consumption through improved operation and maintenance practices.[5] This sounds high, but you may be surprised to learn the percentage of equipment that is not functioning efficiently. In a study sponsored by the California Energy Commission, conducted by David Claridge, of Texas A&M Energy System Laboratory and Hannah Friedman, et al, of the Portland Energy Conservation found:

> ...every building studied had fixes that didn't last. These were overwhelmingly measures that are easily changed. The most problematic and least durable fixes were control strategies like schedules and setpoints that can be modified using a workstation interface.
>
> In the retrocommissioning part of the study, energy savings averaged 41% of the total energy usage decreased by 17% over two years. Although savings decreased, the facilities still saved about 34% of their total energy usage compared to before retrocommissioning. Component failures in two building did not impact comfort but increased energy consumption by $150,000 per year.[4]

Notice the most problematic issues were items people could easily override or modify. Again, human behavior plays a key role in the success of energy initiatives. Notice, too, that malfunctioning equipment that does not affect human comfort is often not repaired. If the room gets too hot, or too cold, a repair is scheduled. If equipment begins using more energy, the problem persists. An effective energy awareness program ensures that everyone knows the value of energy.

System commissioning and maintenance is beyond the scope of this book, but I encourage you to research the issue. You can go beyond the standard approach of repairing systems to the original design intent, and try to optimize facilities for their existing conditions. This can add an additional 5% to 45% savings over normally commissioned systems.[6]

COMPREHENSIVE COMPONENT #3: CONTINUAL MONITORING OF ENERGY INITIATIVES

We have speedometers to show how fast we are driving, gas gauges to alert us to low fuel levels, yet most organizations lack real time gauges to determine energy performance. Building and process operators may have absolutely no understanding of how much energy is used, where it is used, or how their organization compares with similar facilities. However, studies show that monitoring and measuring the results of energy management activities, actually increases energy savings. This is probably intuitive to most managers as measuring and holding personnel accountable for results works in other aspects of business. Once goals are set, we need to track progress and occasionally make midcourse corrections. The best way to accomplish this is to install permanent energy meters and utilize software to turn energy data into useful information.

A United States Department of Energy (DOE) study from the national Renewable Energy Laboratory states: "Case studies have shown that the utility costs can be reduced by 25% or more by identifying energy savings through metering. A less aggressive estimate is approximately 5%."[7] A white paper from Lawrence Berkeley National Laboratory (LBNL) states, "Recent building performance case studies suggests that typical savings of about 15%, and as much as 40% of annual energy use can be gained by compiling, analyzing, and acting upon energy end-use data."[8] I believe the different forecasts are due to the amount of information logged, the type of information logged, and the amount of effort dedicated to acting on the information. The amount of savings a metering system facilitates is proportional to the number of metering points, the software capabilities, and the effectiveness of the actions taken from the resulting information. In any case, the savings are impressive and justify the cost of a metering system.

Even though the meter itself does not directly save energy, the information gleaned from the meter, when acted upon, can produce substantial savings. Information is a powerful tool, what gets measured gets managed. Facility managers should look at utility bill data and metered consumption reports at least monthly, executives at least quarterly. Yet, many mangers never see a utility bill, or interval meter data.

Metering, measurement, and verification are also essential to maximize the savings from energy saving technology and construction projects. Gregory Kats of the U.S. Department of Energy has written several insightful energy documents. In "Energy Efficiency as a Commodity: The Emergence of an Efficiency Secondary Market for Savings in Commercial Buildings[9]," he found:

- Energy efficiency installations undertaken without a method to measure the ongoing results often achieve less than projected savings.

- Installations that use real-time metering to measure savings tend to have higher savings initially and experience savings that stay high.

- When real-time metering is used for a long term duration at multiple locations and projects, savings tend to be more consistent.

Figure 5.3 is a simplified version of a chart from the previously referenced Kats report. Notice how initiatives that are monitored have initial savings equal to or 20% greater than forecasted, and that the results are sustainable. Projects that do not measure the ongoing performance may never achieve their forecasted return and the savings diminish over time. Without good monitoring, most facilities use more energy with time. Goal setting and goal tracking are effective tools in the energy management arena.

In addition to increasing the savings on deployed energy efficient technology, a monitoring system is also an incredible tool to facilitate, boost, and monitor behavioral efforts to reduce energy consumption. It can be used to monitor compliance, track "shut it off" progress to goal, and indicate when messages need reinforcement.

COMPREHENSIVE COMPONENT #4: ENERGY AWARENESS

Energy awareness is, of course, the topic of this book. An energy aware organization understands the cost and importance

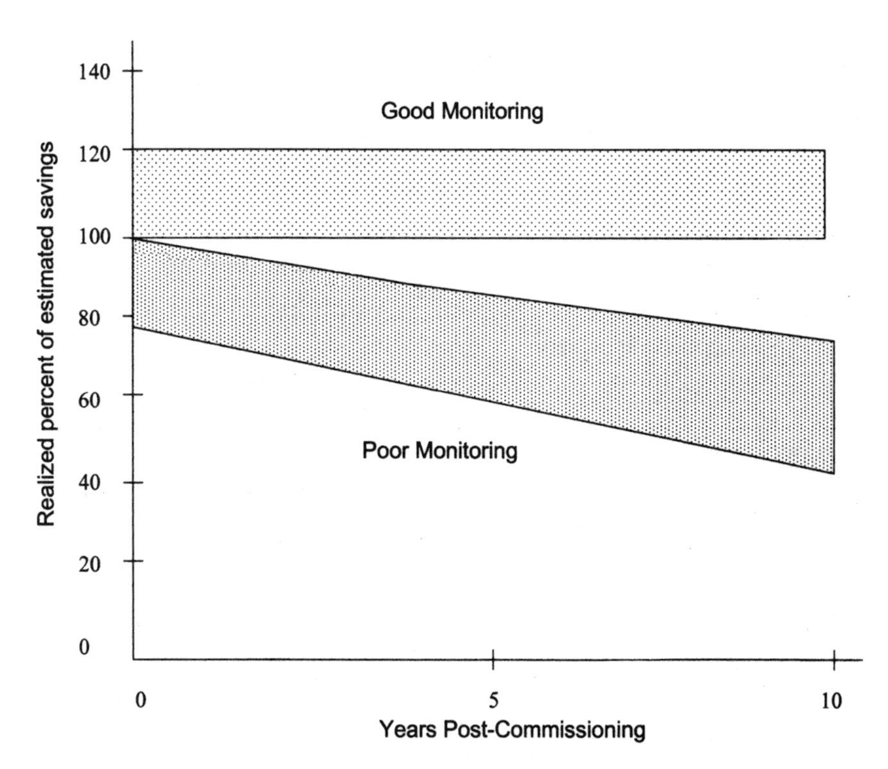

Figure 5.3: The effect of energy monitoring on long-term savings

of energy, tracks and disseminates energy costs to people using equipment and operating processes. People, who can best implement low cost or no cost energy saving initiatives, have the information and support to do so. An effective energy awareness program touches all parts of an organization.

Of course, successful awareness programs are not limited to energy management. We can look at other types of awareness campaigns to understand the potential of such programs. A great example of effective awareness training is the Weirton Steel Corp. Beginning in 1998, they undertook a series of training initiatives, including safety awareness training, and certifying all plant supervisors in the Occupational Safety and Health Administration's (OSHA's) General Industry Standards. As a result, recorded incidents fell 63% from 1997 to 2000.[10]

Obviously, energy awareness programs can use similar methods to educate employees. Carbon Trust, a government-funded agency whose role was to help the United Kingdom move to a low carbon economy, suggests "[Energy] savings of around 15% have been achieved by adopting and maintaining a similar approach to that taken for safety."[11] Energy awareness programs are a proven and essential component of a comprehensive energy management program. Chapter 8 reviews the key components of a successful energy awareness initiative.

SUMMARY

The failure to involve and enlist the cooperation of the organization will negatively affect other seemingly unrelated energy management initiatives. To thoroughly maximize savings, a comprehensive energy management program should include an energy awareness initiative. Alternately, exclusive reliance on an energy awareness program is not optimal. A comprehensive energy management program includes:

1. Technology deployment.
2. Proper commissioning, operation, and maintenance practices.
3. Continual monitoring of energy use.
4. Energy awareness—behavioral and non-technical approaches.

References

1. Natural Resources Canada, Office of Energy Efficiency, "Employee Awareness and the Federal Buildings Initiative, a case study," September 2001. P 1. http://oee.nrcan.gc.ca/communities-government/buildings/federal/case-studies/employee-awareness.cfm?attr=28, (June 22, 2006)
2. Floyd E. Barwig, Iowa Energy center, John M. House, Iowa Energy center, Curtis J. Klaasen, Iowa Energy Center, Morteza M. Ardehali, KN Toosi University of Technology, Theodore F. Smith, The University of Iowa, "The National Building Controls Information Program," *Proceedings from the 2002 ACEEE Summer Study on Energy-Efficiency in Building,* American Council for an Energy-Efficient Economy, Washington, D.C., Vol. 3, 2002, P 1-14.
3. ibid
4. Hannah Friedman, Amanda Potter, and Tudi Haasl of Portland Energy Conservation, Inc. and David Claridge of Texas A&M Energy Systems Laboratory,

"High Performance Commercial Building Systems: Strategies for Improving Persistence of Commissioning Benefits," June 30, 2003.

5. Mary Ann Piette, Tony Sebald, Chris Shockman, Lee Eng Lock, Peter Rumsey, "Development of an Information Monitoring and Diagnostic System" presented at the *Cool Sense national Integrated Chiller Retrofit Forum*, Sept. 23 – 24, 1997, San Francisco, California, LBNL 40512 rev. 2.
6. Charles H. Culp, W. Dan Turner, David E. Claridge, and Jeff S. Haberl, Energy Systems laboratory, Texas A&M University, "Continuous Commissioning in Energy Conservation Programs," http://esl.tamu.edu/programs/morecc.htm, (June 22, 2006)
7. A. Walker, "Advanced Utility Metering, Period of performance: April 23, 2002 – September 22, 2002," *NREL/SR-710-33539*, September 2003, P 25.
8. Frank Olken, Hans-Arno Jacobsen, Chuck McParland, Mary Ann Piette, Mark F. Anderson, "Object Lessons Learned from Distributed System for Remote Building Monitoring and Operation," Lawrence Berkeley National Laboratory 1998, Page 284.
9. Gregory H. Kats, Arthur H. Rosenfeld, Time A. McIntosh, and Scott A. McGaragham, "Energy Efficiency as a Commodity: The Emergence of an Efficiency Secondary Market for Savings in Commercial Buildings."
10. Rachel Madan, "The Human Side of Energy Efficiency: The Value of Training," from the pages of the US Department of Energy, *Energy Matters Newsletter*, summer 2002.
11. A booklet by the UK Department of the Environment, Transportation and the Regions, "Managing and Motivating Staff to Save Energy, Good Practice Guide 84," March 1999, P. 17.

Chapter 6

The Financial Impact of Managing the Behavioral Aspects of Energy Consumption

The largest source of savings arises from turning off equipment when not needed.

—Frank Olken, Hans-Arno Jacobsen, Chuck McParland, Mary Ann Piette, and Mark F. Anderson, "Object Lessons Learned from a Distributed System for Remote Building Monitoring and Operation," Lawrence Berkeley National Laboratory

Far from being an insignificant issue grounded in emotion or ethics, teaching employees to conserve energy is a rational and sound business decision. Intuitively, we can all accept that turning off our numerous unneeded devices will save money. Many of us will even acknowledge that we waste a substantial amount of energy. What may be surprising is how much money companies can save and how easy it is to achieve the savings.

One early pioneer at documenting and monetizing the savings from behavioral programs was the Electric Power Research Institute (EPRI). EPRI is a highly respected nonprofit organization that performs research and development on behalf of the electric utility industry, the utility industry's customers, and society. The *EPRI Journal* was one of the first technical publications I discovered with information on the effectiveness of reducing energy consumption through employee awareness. John Douglas wrote in a 1994 issue of the *EPRI Journal*:

> The effectiveness of reducing energy consumption by introducing improved work practices has been shown through several individ-

ual company programs. IBM, for example, estimates that it saved $17.8 million in 1991 by encouraging employees to turn off equipment after completing tasks and to moderate their use of lighting. Through similar measures, Xerox Canada's corporate headquarters has reduced its energy by 17-23%.[1]

Con Edison and American Express produced a brochure informing employees about the benefits of turning off equipment when not in use. The plan has reduced the facility's electrical bill by more than $700,000 a year.

These were staggering amounts: $17.8 million at IBM, $700,000 at an American Express site, and 17-23% at Xerox Canada's corporate headquarters. In disbelief, I verified the numbers with the EPRI author, John Douglas. He confirmed the savings and the saving methodology, and then referred me to Arthur Kressner of Con Edison. Kressner had worked with American Express on their campaign; he again confirmed the fantastic savings numbers, and wondered why more companies did not implement similar energy savings programs. He enthusiastically discussed the program and shared a brochure[2] American Express had circulated to their employees to help enlist employee cooperation. Some educational excerpts from the brochure:[3]

- Plug all PC components into a master-switched power strip or surge protector. When you turn off the switch on this device at the close of the workday, it ensures that all the components go off.

- Even the smallest daily savings, when added, together, make a big difference.

- Turning off each PC and monitor for 12 hours on nights and weekends and holidays would save $73 per year and would cut emissions by about 3.2 pounds of sulfur dioxide, 2.8 pounds of nitrous oxide, 0.5 pounds of carbon monoxide and 1,800 pounds of carbon dioxide.

These excerpts from the American Express initiative are consistent with other successful education and awareness campaigns that resulted in real and substantial savings. Section II of this book contains many other informational facts and quotes that may be suitable for your energy awareness initiative.

SIMON PROPERTY GROUP

A more recent savings example comes from Simon Property Group, the Real Estate Investment Trust headquartered in Indianapolis, Indiana. In 2004, Simon Property Group implemented a comprehensive strategy to improve energy efficiency. This strategy included the launch of an "Energy Best Practices" program, which challenged managers of enclosed shopping centers to reduce energy costs without affecting customer comfort, safety, or system reliability. The energy management practices reduced electrical usage by 133 million kWhs for 2004 and 2005 combined, as compared to 2003. This represents a 6.8 percent reduction in electrical usage across comparable properties.[4] Using an average price of $0.10 a kWh, this equates to $13.3 million dollars in savings over a two year period, or approximately $6.65 million per year of energy savings.

Not all of the 133 million kWh savings were from low cost, no cost management techniques, but a substantial portion was, according to George Caraghiaur, Vice President of Energy Services for Simon. In a phone conversation, Caraghiaur attributed a large portion of the program's success to increasing the mall managers awareness of the importance of managing energy. Since the mall manager's emphasis is on running a mall and not managing energy, Caraghiaur and his energy staff provided some much needed help and focus. They did this, partially, by pointing out that each shopping center manager is responsible for costs, and energy represents a large portion, about 30%, of their controllable costs. The energy management team in Indianapolis provided the expert knowledge, training, direction, and tracking of results for each shopping center, but the mall managers were on the hook to deliver results. As the local personnel became familiar with their energy consumption and grew comfortable managing energy costs, dramatic savings followed. A little knowledge and accountability can go a long way towards reducing energy costs. Some of the low cost and no cost measures taken include minimizing energy in vacant tenant spaces, using outside air for cooling when possible, and turning off lights. George Caraghiaur and the Simon Property Group illustrated tremendous leadership in energy awareness and continue to reap the financial rewards.

A CANADIAN EXAMPLE

Efficiency at Work is an awareness program designed to save money and promote energy efficiency practices by managing office equipment power loads. It was developed and implemented by the City of Toronto in 2002. The goal is to reduce energy consumption and building operating costs, improve energy security, reliability and help preserve the environment. From the City of Toronto, web site:

> CFB [Canadian Forces Base] Halifax employs approximately 6,000 personnel. Their background research revealed that one computer left on after hours cost $63 per year while a task light left on each night wasted $34 in electrical cost annually. A 10-year, $250,000 employee awareness campaign at the base is estimated to save $600,000 annually, and pay for itself in five months.[5]

The figures are in Canadian dollars, with the $250,000 budget spread over the 10-year period. The expected annual savings from the energy awareness portion was $50,000 a year, the target was easily exceeded and other energy efficiency initiatives were implemented. The information tools used include a traveling project-information booth, fact sheets, project-savings charts, a logo to promote saving energy, refrigerator magnets, posters, newsletters, base newspaper articles, a video, lunchtime information sessions, an "Energy Ideas" contest and a telephone hotline to answer employees' questions concerning the energy performance project.

EXAMPLES FROM THE UNITED KINGDOM

Another well-respected source of documented energy savings can be found at Carbon Trust, an independent and nonprofit government funded organization that works with companies in the United Kingdom to cut emissions of carbon dioxide and reduce climate change. Carbon Trust has documented the following success stories.

Solutia

Solutia is an excellent applied chemistry company that achieved savings of £350,000 through no-cost & low-cost solutions.[6] One press release stated:

> Solutia UK Ltd is a chemical manufacturing company based in Newport, Gwent. It has 7 manufacturing plants on its 320 acre site, 230 employees and operates 24 hours a day, 365 days a year. The company's annual business energy bill is in the region of £2.5 million.
>
> The company adopted a wide range of low and no cost energy efficiency measures, resulting in a 13% decrease in absolute energy, equivalent to £350,000, against a real increase in production of 7%. This has mainly been achieved through an employee awareness campaign within the organization.[7]

Not only is £350,000 a lot of money in real terms, but it is an impressive percentage of the £2,500,000 annual bill.

Hewlett-Packard

Another good example comes from Digital Equipment Company, a computer company that was acquired by Compaq, which then subsequently merged with Hewlett-Packard. Bottom

Line Benefits:

> Digital has calculated that annual electricity savings worth around £100,000 have resulted from the switch off at night campaign. Internal costs, equating to the production of the posters and network programming for the switch off message, have been estimated as only £5000.[8]

Digital found that the attitude of the staff can have a significant effect on the energy bill. The energy manager used different training and motivational material for the staff engaged in facilities management and the office staff. For example: The facilities management staff was given monthly graphs of daily energy use, while the office staff received messages when they logged out of e-mail, stating the amount of Digital's energy bill and asking users to turn off their computers. Colorful poster and articles in the company's in-house journal were used to help reinforce the messages.[9]

Rover Group

Rover is an automobile manufacturer that saved over £1,000,000 the first year of their awareness campaign. From Action Energy:

> Awareness programs are a very low risk option in energy management. At Longbridge there have been many simple, no-cost actions taken. Just one of these, the rescheduling of one plant item, produced savings of over £30,000—more than four times the total cost of the program.
>
> The savings achieved at Longbridge during the awareness program were in excess of £1M. The cost of producing the newsletter and other promotional literature was £7,200. This results in a negligible simple payback period.[10]

Some steps Rover took:

- They formed a site energy group. The team was lead by the manufacturing engineering director. He had a core of

15 members who represented various activities including engineering, manufacturing, personnel and finance. Team membership was flexible and other people were involved as required.

- They created a publicity campaign and produced a six page, full color special newsletter that was mailed to employees and their families.

- They started reporting energy performance against targets. A weekly report informed all employees of energy usage and how they were tracking to goals.[11]

Iceland Foods

Iceland Foods, a grocery store chain, launched the motivation and awareness campaign to reduce electricity costs in their stores by raising staff awareness of energy practices. They saved over £200,000 through their awareness program. From Action Energy:

Savings Achieved

Energy savings worth £200,000 in 1990. As a result of the awareness campaign an additional £15,000/year is saved by lowering maintenance costs and the avoidance of capital expenditure by extending the life of equipment.

Investment Cost

£20,000 on design and print of posters and stickers, in-house training initiatives and regular distribution of energy information. Payback Period Less than 6 weeks.[12]

Iceland transmitted energy cost information to the stores via the existing communication structure for feedback and performance monitoring. All stores received advice and training seminars on energy awareness. Posters and stickers helped maintain momentum. A £1,000 prize was awarded to the best region, while small prizes were given for the best energy saving ideas. The cost of implementing the program, including the £1,000 prize, was recovered through energy savings in only six weeks, clearly a good investment.[13]

University of East Anglia

Universities are also successfully reducing energy costs through energy awareness programs. The University of East Anglia is a good example:

> *Results*
>
> The monitoring results showed that, although the electricity use of the residences was already very low, the 3058 residential students have achieved annual energy savings of 168,600 kWh through simple good housekeeping practices in areas where they have direct control. The saving is equivalent to about 8% of controllable electricity usage and provided a reduction in CO_2 emissions equivalent to 32 tonnes of carbon. If similar savings to those of UEA could be replicated across the whole higher and further education sector [UK], then annual CO_2 production would be reduced by the equivalent of 2230 tonnes of carbon.
>
> The cost saving achieved through the energy awareness campaign was £6200 of which 25% was returned to the students, via the SU [Student Union], for spending as they pleased.[14]

Pamphlets outlining the program and giving specific energy-saving tips were distributed to each university resident during their first month of residency. A series of posters was displayed in communal areas of each residence. The posters were updated approximately every two months to keep the message fresh. Student volunteers and cleaning staff distributed informational materials during regular house rounds. Student volunteers from the University's Society for Environmental Action also talked to residents explaining the campaign and its benefits. In addition, other initiatives, such as energy tip bookmarks, newspaper articles, and a mobile energy center, were used to target the whole student population.[15]

MORE EXAMPLES FROM THE UNITED STATES

Let's get back to a few examples from the United States. With less than 5% of the world's population, the United States consumes one quarter of the world's energy, spending about $300 billion each year on just electrical energy. The state of California initiated "Flex Your Power," a statewide energy efficiency marketing and outreach campaign in 2001. The program began after an unfortunate combination of events caused power outages as the state was deregulating its electrical utility industry. Flex Your Power is an outstanding program, which has received national and international recognition, including an ENERGY STAR Award for excellence.

There is a high likelihood that the sheer magnitude of the California crisis increased the savings of awareness programs in California by a statistically significant amount. There were persistent threats of power outages, and residents were beset with rumors of highly escalating energy costs in both their homes and workplaces. Thus, in this case, we must acknowledge that a portion of the behavioral efforts may have been motivated by personal fear of blackouts and fear of higher home energy bills. This is not to take anything from the program; it was extremely well run and essential. As a matter of fact, Flex Your Power is widely credited with preventing blackouts during the crisis. That said, there are some awareness program lessons and examples worth reviewing:

Thomas Properties Group LLC

In 2001, Thomas Properties utilized a simple, repeatable, and low cost approach to reduce energy costs and increased tenant satisfaction at the same time. From Flex Your Power:

> Thomas Properties Group in Sacramento initiated a new janitorial cleaning program for its building. Previously, janitors cleaned from 6 p.m. to 2 a.m. with 13 floors fully lit. Under a new cleaning and lighting program, janitors started work at 11:30 a.m., often using only core lighting.
>
> The new work schedule saved Thomas Properties $50,000, or 4.5% of the energy cost annually.[16]

As a side benefit of this new schedule, the janitorial staff had weekends off and received 70 percent fewer complaints from tenants about the quality of the cleaning service. This was a win-win for the tenants, the cleaning staff, environment, and the energy bill.

One U.S. Department of Energy (DOE) presentation indicated that an average energy savings potential of $0.01 a square foot in commercial buildings, or 5% of overall electrical use is attainable through day cleaning. There is an optional small investment in tools, such as battery powered floor buffers so that there are no cords to trip over and silent vacuums so as not to disturb tenants.

Verizon

Verizon expanded their energy awareness initiative to locations outside of California, saving $17 million dollars. This is a staggering number, with most of the savings realized not in black out prone California. From Flex Your Power, Business Guide 3: *Target Business Employees for Energy Conservation in the Workplace*:

> Verizon's employee awareness efforts cost less than $5,000 to implement—but yielded savings of $750,000 and 10 million kWh in California alone.
>
> Nationwide employee action contributed to $17 million in savings and 11.5 percent reduction in energy use in 82 administrative target locations (compared with 2 percent in unoccupied buildings) between January and September 2001.[17]

In another Flex Your Power Business Case Study: Verizon Communications, a more in-depth look at Verizon shows they, like most of the other companies in this chapter, used a balanced approach to energy management, using both employee awareness and technology:

> The estimated annual energy and financial savings [in California] were as follows:

- Lighting and A/C upgrades: 44.8 million kWh and $1.7 million in avoided energy cost.

- Delamping: 4.9 million kWh and $438,000 in avoided costs.

- Energy communications/awareness: 10 Million kWh and $750,000 in avoided energy costs.

Verizon found:

Communications efforts are difficult to quantify, but it was clear to Verizon that they made a big difference. As employees became more involved and vocal about energy issues, the conservation message spread.

Verizon developed a company wide "Energy Champions" program, which selected key individuals to take responsibility to educate and motivate other co-workers at each location. The awareness campaign consisted of frequent progress updates to staff, energy letters from senior VPs, an energy website, an energy newsletter, large billboards with messages, and an energy hotline to gather suggestions.[18]

Lockheed Martin

Lockheed Martin made its vice presidents responsible for the energy use in groups of buildings (called a Chunker). From Flex Your Power:

Lockheed Martin Missiles and Space credited a large portion of its 2001 energy savings to its low-cost employee awareness and incentive programs. Audits conducted after the launch of employee awareness programs and Chunkers revealed that fewer lights and office equipment were being left on compared with data from previous audits.

- Employee action contributed to: less wasted energy; $4,800,000 in cost avoidance/financial savings annually; 75 MW of energy use reductions/savings daily; and 13.6 percent electricity savings and 17.3 percent natural gas savings in 2001.

- The energy fair cost $1,300 for posters, coffee mugs and other items and the conservation contest cost approximately $200 in rewards.[19]

Unisys

Let's review the 2001 energy awareness program at the Unisys Corporation. Another example from Flex Your Power:

> Unisys' Summer Survivor Program cost $11,400 for incentives (lunches, buttons and logo flashlights) and $2,400 for personal fans. The conservation actions produced remarkable results: 22 percent power reduction (1.8 million kWh) and $263,250 in energy costs saved over the five-month summer period and an overall 19 percent power reduction for 2001.[20]

Unisys, capitalizing on the TV show "Survivor," named their awareness program, "Summer Survivor." They awarded $5 lunch certificates to employees for their saving efforts. Additionally, they provided free lunches to all 1,000 employees during the special "Energy Conservation Results" event. Personal fans were provided to maintain employee comfort when it was necessary to cycle the air conditioning on and off. This was above and beyond the normal energy awareness campaign, but one must remember that at this time California faced a severe situation.[21]

FEDERAL ENERGY MANAGEMENT PROGRAM (FEMP) STUDIES

In addition to the savings examples we already discussed, FEMP conducted two pilot projects to demonstrate that behavior-based programs in a residential setting can help reduce energy use and expenditures. The programs targeted families in military housing. Military residents do not pay their own utility bills, so the reductions in energy use resulted purely from behavioral changes, and not financial gain. At Fort Lewis, Washington, the pilot program ran for one year. The total energy saving was 10% on a weather-corrected basis and totaled over $130,000 for the year. The other study was at the Marine Corps Air Station in Yuma, Arizona. It ran for only the three months of the summer cooling season and energy use dipped 13% in the last month, leaving the housing manager with a $50,000 surplus at year-end![22]

SUMMARY

While your actual energy and dollar savings will depend on the type and amount of equipment you have, your regional power costs, and effectiveness of your efforts, it is clear from the examples in this chapter that the cost of acting on behalf of an energy awareness campaign is low, but the consequences of not acting are high. Educated and motivated individuals can produce real savings with a modest investment. Once more, consider the reviewed examples of yearly savings:

- American Express, New York facility—$700,000.
- Canadian Forces Base, Halifax—$600,000 (Canadian dollars).
- Digital Equipment, UK –£100,000
- IBM—$17,800,000.
- Iceland Foods, UK—£200,000 of energy and £15,000 on maintenance costs.
- Lockheed Martin—Employee action contributed to $4,800,000 in cost avoidance annually.
- Rover, Longbridge—£1,000,000.
- Simon Property Group—A substantial portion of 133 million kWhs for 2004 and 2005 combined, as compared to 2003. This represents a 6.8 percent reduction in electrical usage across comparable properties. At a rough average price of $0.10 a kWh this equates $13.3 million dollars in savings over a two year period, or approximately $6.65 million per year of energy savings.
- Solutia—£350,000 was mainly achieved through an employee awareness campaign targeted at employees.

- Thomas Properties Group, LLC, Sacramento—$50,000.
- Unisys, California—$263,250 in five months and 19 percent power reduction for 2001.
- University of East Anglia—£6,200.
- Verizon, California—$750,000.
- Verizon, nationwide—$17,000,000 (partial year savings between January and September 2001).
- Xerox Canadian headquarters—Reduced usage 17-23%.

These savings examples should provide you with the motivation to think about implementing an energy awareness program in your organization. Saving money should be incentive alone to take action. I realize that action and decisions require a greater justification than inaction, or the failure to decide. If you need more justification, the environmental consequences of electrical consumption in the next chapter may provide the additional motivation to launch an awareness program.

References

1. John Douglas, "The Energy Efficient Office," *Electric Power Research Institute EPRI Journal*, July/August 1994, P. 18.
2. A brochure by American Express and Con Edison, "Bright Ideas For Saving Energy At American Express While Protecting Our Environment," October 1991.
3. ibid
4. Simon Property Group, "Form 10-K annual report pursuant to section 13 or 15 (d) of the securities exchange act of 1934, for fiscal year ending December 31, 2005," P 35. Available through Simon.com (June 22, 2006).
5. City of Toronto: Energy Efficiency web site, About E3@Work, www.city.toronto.on.ca/energy/e3atwork/about.htm, July 2005, (June 22, 2006)
6. 1 UK pound sterling (£) has fluctuated somewhere between $1.40 to $1.95 U.S over the past ten years.
7. "Business and Environment Action Plan for Wales—First Annual Report, 2003-04," Page 5; and "Energy with the gloves off," The Manufacturer, September 5, 2003.
8. "Energy management—awareness and motivation, Digital Equipment Company Limited, a Good Practice Case Study 341," September 1996. A Carbon Trust Good Practice Case Study, Publication ID GPCS341
9. ibid
10. "Energy saved by raising employees' awareness, A Good Practice Case Study 214," August 1996. Publication ID GPCS214 available at carbontrust.com (June 22, 2006)
11. ibid
12. "Energy efficiency motivation campaign in a multi-site organization, A Good Practice Case Study 182," March, 1998. Publication ID GPCS182 available at carbontrust.com (June 22, 2006)

13. ibid
14. Carbon Trust, "Student energy awareness scheme—University of East Anglia," A Good Practice Case Study 367. Carbon Trust Publication ID GPCS367
15. ibid
16. State of California, "Flex Your Power Energy Conservation and Efficiency Campaign 2001-2002, 2003," P. 7. and State of California, "Flex Your Power Business case study; Thomas properties Group."
17. State of California, "Business Guide 3: Target Business Employees for Energy Conservation in the Workplace," A guide from the California Flex Your Power Initiative. P 9.
18. State of California, "Business Case Study: Verizon Communications," A California Flex Your Power Business Case Study.
19. State of California, "Business Guide 3: Target Business Employees for Energy Conservation in the Workplace," A guide from the California Flex Your Power Initiative. P 4, 8.
20. State of California, "Business Case Study: Unisys Corporation for Flex your Power Campaign and State of California, Saving Energy. It's a way of Life, Unisys, Mission Viejo," from California Flex you Power Campaign.
21. ibid
22. U.S. Department of Energy, "Creating an Energy Awareness Program, A Handbook for Federal Energy Managers," P. 2.

Chapter 7

The Environmental Impact of Energy Consumption

The biggest challenges that Humankind is facing today in industrialized nations are of two types: awareness on the part of people about the ecological impacts of their consumption patterns, and the lack of the ways and means to extricate themselves from their addiction to this ecologically hostile consumption pattern.

—Gopi Upreti "Environmental Conservation and Sustainable Development Require a New Development Approach," *Environmental Conservation*

Knowledge of the environmental consequences of energy consumption can be a powerful motivator to elicit and gain "energy-saving" support from individuals. A good example of evoking this environmental sentiment is readily apparent at hotels that ask guests, through internal signs and plaques in hotel rooms, to make ecologically sound choices. In fact, a large number of hotels have guest energy awareness programs of some sort. On the following two pages are samples of the messaging on cards strategically placed in various hotel rooms.

It is now a common practice at many hotels to offer hotel guests the opportunity to re-use towels and not have bed sheets changed everyday. Some hotels also encourage guests to turn off lights and air conditioning when the room is vacant. Many hotels provide recycling bins in each hotel room for newspapers and other items. One study reported, surprisingly, 95% of hotel guest were happy to see the conservation efforts, and that individual

Dear Guest,

We are committed to energy and resource conservation.

Energy conservation helps reduce the effects of global warming, acid rain, and smog, and protects our natural resources. Water conservation is a vital concern as well.

Our standard is to change your bed linen every third day and at check-out. If you would like your sheets changed during your stay, we will gladly accommodate you. Just place this card on your pillow or call housekeeping.

In addition to the above, we've also pioneered recycling programs and installed water saving devices and energy saving lighting throughout the hotel.

Thank you

A member of the Hilton family of hotels

At Wyndham, we know that many of our guests are interested in helping to protect the natural resources of our planet. That's why we've instituted a new program- EarthSmart@ that can reduce the amount of water, energy and detergents that we use daily.

You can help by using your towels more than once. Just hang your towels back on your bathroom rack if you would like to reuse them. Otherwise, simply leave them on the floor and our housekeeping staff will provide you with fresh towels.

Thanks for helping us make a difference!

Wyndham Hotels & resorts

> *Make a world of difference—help us conserve.*
>
> *Linens*
> *We will make your bed every day. Your linens will be changed only when this card is placed on the bed in the morning.*
>
> *Towels*
> *Leave towels you wish to reuse hung up or on the rack. Towels you leave on the floor will be washed.*
>
> *Together we can save millions of gallons of water, chlorine and detergents.*
>
> *Sheraton Hotels and Resorts*

hotels were saving about $5,000 to $100,000 per year, depending on the size of the hotel and average length of the guests' stay.[1] Having guests reuse towels and sheets also lowers housekeeping labor costs. The labor savings (of not needing to completely remake beds and shuffle towels around) can exceed the actual energy cost savings, a side benefit of protecting the environment. Remember the split incentive pitfall from Chapter 4? In this case, hotel guests have no financial incentive to save water or energy, as all the cost savings go to the hotel owner, but the desire to help the environment proves to be a strong motivator that can counteract this obstacle.

Organizations can leverage this environmental goodwill to reduce energy costs and their associated emissions. The means that environmental education should be a key part of all energy awareness programs. The goal of environmental education is to motivate people to act environmentally responsible, or in this case, to reduce energy costs. Let's examine the impact of electrical power generation on our environment.

EFFECT OF ELECTRICAL POWER ON THE ENVIRONMENT

You may be surprised to learn that electric power plants are the largest source of many of the pollutants that contribute to global warming, acid rain, and mercury poisoning in lakes and rivers in the United States. It is easy to forget that somewhere a fire is burning, generating steam that turns a generator, sending toxic emissions through a smoke stack. Powering our lights, computers, and air conditioning systems creates substantial amounts of air pollution.

The major environmental impact of electrical power comes from burning coal. Figure 7.1[2] indicates the percentage of electrical energy generated by coal in key regions across the globe. As the chart shows, much of the world relies on coal to generate electrical power. The United States is widely considered to be the Saudi Arabia of coal with some of the largest known deposits in the world. Coals' relative abundance continues to make it financially attractive versus other fuels, such as natural gas and oil. Given the abundance of coal in the United States and its relatively low cost, you may expect to see a larger percentage of coal fired power. Various environmental and permitting regulations have temporally kept the coal-fired power percentage at a low level in the United States. China is both the largest consumer and producer of coal (for all purposes), consuming 28% of the world total.

ELECTRICAL POWER IS LARGEST HUMAN-CAUSED SOURCE OF MERCURY EMISSIONS TO THE AIR IN THE UNITED STATES

Coal burning electrical power plants are a large source of mercury emissions, and the largest human caused source of mercury emissions to the air in the United States. Approximately 75 tons of mercury is found in the coal delivered to U.S. power plants each year and about two thirds of this mercury is emitted into the

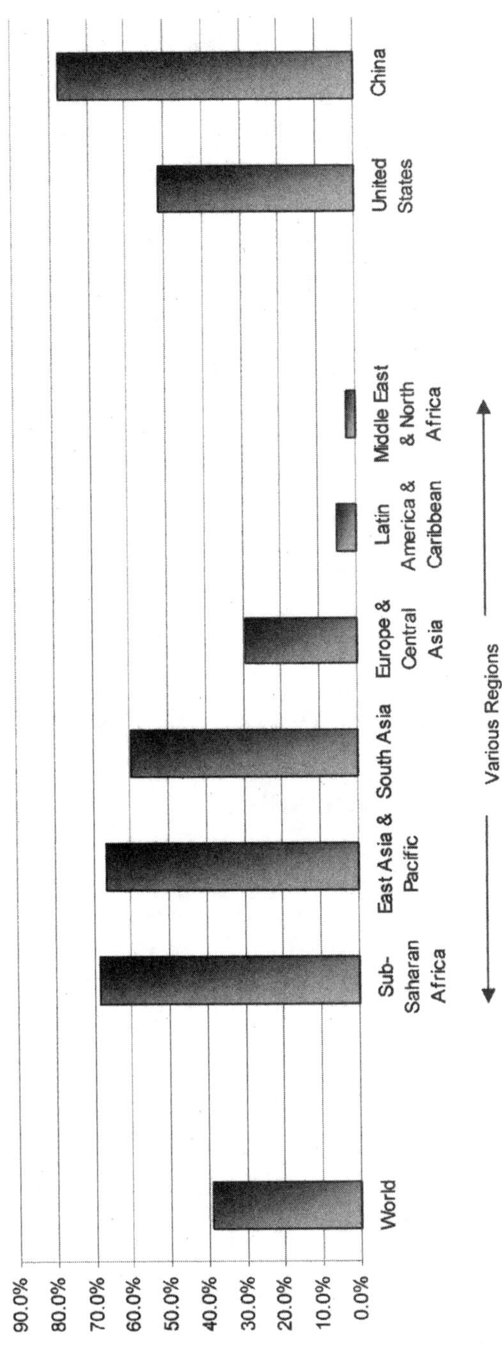

Figure 7.1: Percent of electrical energy generated by coal in various parts of the world.

air, resulting in about 50 tons being emitted annually.[3]

The mercury in the air eventually settles onto the land or into the water. Once deposited, certain microorganisms can change it into methylmercury, a highly toxic substance that builds up in fish, shellfish, and animals that eat fish. Fish and shellfish are the main sources of methylmercury exposure to humans. Mercury exposure at high levels is very damaging. Mercury compounds can attack all body systems including the brain, heart, kidneys, lungs, and immune system of people. Methylmercury is extremely dangerous to pregnant women. Methylmercury, usually from contaminated food, can cause mental retardation, cerebral palsy, seizures, and eye and hearing damage in a fetus because of the mother's exposure.

You do not need to live close to an electrical power plant to be affected by methylmercury. The mercury in the air and the contaminated food we eat are often transported great distances. Fortunately, we can reduce the amount of mercury in our environment simply by turning off unneeded devices.

ELECTRIC POWER GENERATION IS THE LARGEST SINGLE SOURCE OF SULFUR DIOXIDE EMISSIONS IN THE U.S.

Sulfur is present in coal as an impurity. It reacts with air when coal is burned and forms sulfur dioxide. Electric power generation is by far the largest single source of sulfur dioxide emissions in the U.S., accounting for approximately 67 percent of the total sulfur dioxide emissions nationwide.[4] Despite improved emission controls on electrical power plants, we have seen a slight increase recently resulting from increased production of electrical energy by coal-fired and oil-fired units that emit much more sulfur dioxide than natural gas units. This trend may continue as natural gas prices rise.

Even brief exposure to relatively low levels of sulfur dioxide has been repeatedly shown to trigger attacks in people with asthma. Sulfur dioxide also causes acid rain and eye damage.[5]

THE UNITED STATES ELECTRIC POWER INDUSTRY ACCOUNTS FOR APPROXIMATELY 22 PERCENT OF THE TOTAL ANNUAL NITROGEN OXIDE EMISSIONS

Nitrogen oxide is formed when any fossil fuel is burned. Power plants are the source of approximately 22 percent of the total annual nitrogen oxide emissions, a major component of ozone smog and fine particulate matter. Smog is formed when nitrogen oxides (NO_x) from electrical power plants and cars mix with other chemicals in the air in the presence of sunlight. Nitrogen oxide causes headaches, reduces lung function (in anyone exposed), and exacerbates asthma. It contributes to formation of acid rain, reacts to form toxic chemicals, and contributes to global warming. It effects the health of millions of people.[6]

CARBON DIOXIDE (CO_2)

Power plants are responsible for more than a third of the U.S. CO_2 emissions. In 2003, the electric power plants in the United States created more CO_2, 2,279.3 million metric tons, than the transportation sector at 1,874.7 million metric tons.[7] CO_2 emissions from the United States electric power sector have grown by 27.5 percent since 1990. Most people don't stop and reflect that the smoke stack at the electrical power plant may produce more emissions powering their home or office than driving their car does, again because the smoke stack is "out of sight and out of mind."

Removing CO_2 from smoke stack emissions is not as easy or inexpensive as the removal of other pollutants. Cost estimates are 2¢ or more per kWh; consequently, CO_2 emissions continue to rise faster than other pollutants.

Many scientists consider CO_2 to be the major source of global warming. The global warming theory states that if the amount of energy (heat) transmitted to the earth by the sun is equal to the amount of energy (heat) re-radiated back into space, the earth's temperature will remain roughly constant; however, many gases in the Earth's atmosphere absorb the energy (heat) re-radiated from the surface, trapping the heat in the lower atmosphere. You are probably familiar with this effect, experiencing colder nights with clear skies and warmer nights with thick cloud cover; invisible CO_2 can have a similar effect to that of clouds.

Therefore, scientists are concerned that CO_2 concentrations in the earth's atmosphere are increasing every year and suggest that increasing concentrations of greenhouse gases will make the earth warmer. Some signs of global warming may appear as a paradox. For example, the initial signs of global warming may be cooler summers and winters as portions of the oceans cool temporarily from melting polar ice.

The "Keeling Curve," Figure 7.2[8], is a graph of CO_2 measurements from Mauna Loa, Hawaii, an ideal location for measuring CO_2. The Mauna Loa atmospheric CO_2 measurements are the longest continuous record of real-time atmospheric CO_2 concentrations available. A new peak on the graph occurs every winter. Each new valley occurs in the summer as trees and plants in the northern hemisphere absorb CO_2 during their growth cycle. The record shows a 19.4% increase in the mean annual concentration of atmospheric CO_2, from 315.98 parts per million by volume (ppmv) of dry air in 1959 to 377.38 ppmv in 2004.[9] The overall rise in CO_2 levels are attributable to burning fossil fuels such as oil and coal. Notice how the CO_2 levels keep rising.

The trend is expected to accelerate as the world population grows and living standards improve globally. In 2005, there were about 6.3 billion people in the world with forecast of 9 billion by 2050. Most of the population growth is occurring in less developed countries that are striving to, and acquiring, more electrical power, cars, and other energy intensive habits each day. For example, China currently has on average 7 cars per 1,000 people whereas the United States has 481, France has 491, and Germany 516 cars per 1,000 people, but China is ramping up car production for domestic consumption as their economy grows.[10] As the need for appliances such as air conditioning, televisions, computers, and phone systems increases, the demand for electrical power

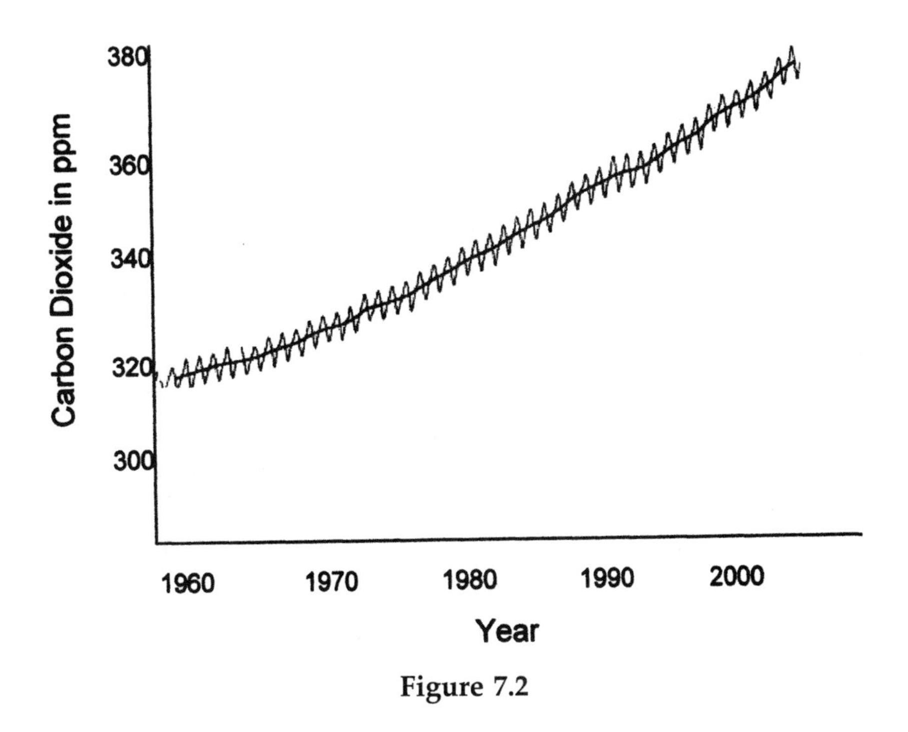

Figure 7.2

may rise to four times the 2002 level by 2005.[11] Consequently, it is easy to forecast a continual increase in the concentration of atmospheric CO_2 if business continues as usual.

If some scientists are correct, a warming of 1.8°F to 7.2°F in global mean temperatures is anticipated over this century. This will cause major changes in the World. Our climate is the single most important factor determining the distributions of major vegetation types and individual species. Changes could include, among other things:

- The spread of warmer climate infectious diseases such as malaria dengue, tick borne encephalitis, Leishmaniasis, Lyme borreliosis (or Lyme disease), West Nile fever and Hantaviruses.

- Loss of fresh water fisheries, such as much of the salmon habitat.

- Changes in vegetation patterns and animal habitats possibly leading to extinctions.
- Drought and heat waves.
- Larger storms creating floods, hurricanes, cyclones and tornados.
- A substantial rise in sea levels.

There are alternate theories that global warming will lead to cooler temperatures. The concept is that global warming will increase evaporation of ocean water causing more rain and snow. The increased snow cover will build and reflect the sunlight that would otherwise warm the earth and create another ice age.

One thing is certain, carbon dioxide levels are at all time recorded highs and climbing. Many countries are adopting stricter standards on emissions hoping to avoid or postpone the consequences of a new atmospheric mixture.

NEEDLESS ENERGY CONSUMPTION PROBABLY KILLS PEOPLE

It is sobering to consider that the light left on needlessly may have contributed to the demise of another person, but one American Lung Association study indicates that air pollution from power plants might be responsible for 30,000 premature deaths each year in the United States. From the American Lung Association web site:

Air Pollution from Power Plants Responsible for 30,000 Premature Deaths Each Year in U.S.

This analysis by Abt Associates used EPA-approved emissions and air quality modeling techniques to forecast ambient air quality in 2007, assuming full implementation of the Clean Air Act's acid rain control program, and the EPA's 1999 "NO_x State Implementation Plan (SIP) call." Analysts then applied risk functions derived from epidemiological studies to estimate health impacts of power plant emissions in the U.S.

The focus of the study was on gaseous emissions of sulfur dioxide and nitrogen oxides that are converted in the atmosphere to fine particle sulfates and nitrates. The analysis estimated that 30,100 deaths may be attributed to power plant emissions each year. In addition, power plant emissions cause 20,100 hospitalizations for respiratory and cardiovascular causes, more than 7,000 asthma-related emergency room visits, 18,600 cases of chronic bronchitis, 600,000 asthma attacks, over 5 million lost work days, and over 26 million minor restricted activity days. Reductions in emissions from uncontrolled power plants could substantially reduce the adverse health effects.[12]

CALCULATING UNITED STATES EMISSIONS

The amount of emissions individuals and businesses are responsible for varies depending on the type of fuel the local power plant burns, the extent of pollution controls it utilizes, and, of course, the amount of energy consumed. Electrical power sources such as hydroelectric power from dams, solar, and wind turbines don't burn fossil fuels or create pollution as they generate electrical energy. If you are lucky enough to live in an area that utilizes clean power, please realize that you can still reduce emissions by conserving power. We are basically on a large energy grid; the energy you save is exported and used by other customers on the grid, often in lieu of burning fossil fuels that are more expensive.

The average pollution emissions table (Table 7.1)[13] will allow you to calculate the emissions from electrical energy consumption in the United States. Ideally, you could contact your utility and get the exact data; however, these averages should provide a good representation of your pollution effects. When calculating the emissions, consider using the savings ratios from the preceding chapter, and ask yourself what percentage of the pollution is avoidable? An average company can reduce energy consumption 10 to 15% through low-cost, no-cost, behavioral management techniques.

Table 7.1: U.S. average pollution emissions per kWh. Source: Energy Information Administration, *Updated State- and Regional-level Greenhouse Gas Emission Factors for Electricity* (March 2005).

Region/State	CO_2 lbs/kW	CH_4 lbs/MWh	N_2O lbs/MWh
New England	**0.98**	**0.0207**	**0.0146**
Connecticut	0.94	0.0174	0.012
Maine	0.85	0.0565	0.027
Massachusetts	1.28	0.0174	0.0159
New Hampshire	0.68	0.0172	0.0141
Rhode Island	1.05	0.0068	0.0047
Vermont	0.03	0.0096	0.0039
Mid Atlantic	**1.04**	**0.0093**	**0.0145**
New Jersey	0.71	0.0077	0.0079
New York	0.86	0.0081	0.0089
Pennsylvania	1.26	0.0107	0.0203
East-North Central	**1.63**	**0.0123**	**0.0257**
Illinois	1.16	0.0082	0.018
Indiana	2.08	0.0143	0.0323
Michigan	1.58	0.0146	0.025
Ohio	1.8	0.013	0.0288
Wisconsin	1.64	0.0138	0.026
West-North Central	**1.73**	**0.0127**	**0.0269**
Iowa	1.88	0.0138	0.0298
Kansas	1.68	0.0112	0.0254
Minnesota	1.52	0.0157	0.0247
Missouri	1.84	0.0126	0.0288
Nebraska	1.4	0.0095	0.0219
North Dakota	2.24	0.0147	0.0339
South Dakota	0.8	0.0053	0.0121
Texas	1.46	0.0077	0.0146

Region/State	CO_2 lbs/kW	CH_4 lbs/MWh	N_2O lbs/MWh
Mountain	**1.56**	**0.0108**	**0.0236**
Arizona	1.05	0.0068	0.0154
Colorado	1.93	0.0127	0.0289
Idaho	0.03	0.008	0.0033
Montana	1.43	0.0108	0.0227
Nevada	1.52	0.009	0.0195
New Mexico	2.02	0.0131	0.0296
Utah	1.93	0.0134	0.0308
Wyoming	2.15	0.0147	0.0338
Pacific Contiguous	**0.45**	**0.0053**	**0.0037**

Table 7.1: (*Continued*)

Region/State	CO_2 lbs/kW	CH_4 lbs/MWh	N_2O lbs/MWh
California	0.61	0.0067	0.0037
Oregon	0.28	0.0033	0.0034
Washington	0.25	0.0037	0.004
Pacific Non-contiguous	**1.56**	**0.0161**	**0.0149**
Alaska	1.38	0.0068	0.0089
Hawaii	1.66	0.0214	0.0183
Region/State	CO_2 lbs/kW	CH_4 lbs/MWh	N_2O lbs/MWh
South Atlantic	**1.35**	**0.0127**	**0.0207**
Delaware	1.83	0.0123	0.0227
Florida	1.39	0.015	0.018
Georgia	1.37	0.0129	0.0226
Maryland	1.37	0.0118	0.0206
North Carolina	1.24	0.0105	0.0203
South Carolina	0.83	0.0091	0.0145
Virginia	1.16	0.0137	0.0192
West Virginia	1.98	0.0137	0.0316
East-South Central	**1.49**	**0.0128**	**0.024**
Alabama	1.31	0.0137	0.0223
Kentucky	2.01	0.014	0.0321
Mississippi	1.29	0.0132	0.0165
Tennessee	1.3	0.0105	0.0212
West-South Central	**1.43**	**0.0087**	**0.0153**
Arkansas	1.29	0.0125	0.0203
Louisiana	1.18	0.0094	0.0112
Oklahoma	1.72	0.011	0.0223
Texas	1.46	0.0077	0.0146
Region/State	CO_2 lbs/kW	CH_4 lbs/MWh	N_2O lbs/MWh
U.S. Average	**1.34**	**0.0111**	**0.0192**

To use the table, locate your energy bill and determine the total number of kWh consumed. Multiply the kWh consumption by the number in the second column that corresponds with your region or state. This number represents the total carbon dioxide emissions in pounds from your energy consumption. For methane (CH_4) or Nitrous Oxide (N_2O) use the same process and then divide by 1000 to change the MWh to kWh. Methane and Nitrous Oxide are important greenhouse gases that can remain in the atmosphere for long periods. Methane may be up to 20 times more effective at trapping heat in the atmosphere than carbon dioxide over a 100-year time period.[14] Nitrous oxide is estimated to be 310 times more effective at trapping heat in the atmosphere than carbon dioxide over a 100-year time period.[15] The emission figures do not include the energy used to mine and transfer the coal or other resource to the power plant, simply the pollutants emitted from actually burning the resource.

Examples using the emission table: If your home consumed 2,000 kWh last month and you live in New England the math will look like this:

> 2,000 kWh * .98 lbs/kWh = 1,960 lbs of Carbon Dioxide emitted into the atmosphere as the result of electrical energy use.

> 2,000 kWh * 1MWh/1000 kWh * 0.0146 lbs/MWh = 0.0292 pounds of Nitrous Oxide emitted into the atmosphere as the result of electrical energy use.

If you want to determine the emission impact of turning off specific devices, such as a computer or light. We first need to calculate the total kWhs saved. For example, a 100 watt light bulb turned off for 10 extra hours a day saves 1 kWh (100 W * 10 H = 1,000 WH, or 1 kWh). Over a year this would total 1kWh/day * 365 days a year = 365 kWh. Then, for Connecticut, the calculation is .98 lbs CO_2/kWh * 365 kWh = 357.7 pounds of carbon dioxide. In the summer, and possibly in the winter, this would be a conservative estimate, as additional energy is required to cool the building to compensate for the heat generated by the light bulb.

A word of caution, computers list the maximum power (W) they may consume, not the average power they consumed. Remember that a watt is an instantaneous reading similar to the speed on a speedometer. A car speedometer may go up to 180 miles per hour, that does not mean the car will go that fast, or average that speed. Similarly, the 600W label on a PC indicates a 600W power supply, which means the computer power supply is capable of delivering 600W, but it may never deliver 600W as it is oversized to accommodate potential or future accessories. A workplace desktop PC may draw approximately 30-135 W for the processor, 10-50W for the motherboard and memory, the graphics cards can add 15-130W, maybe 10W for the hard drive, fans and power supply losses also add up. A typically desktop business computer might be in the 100- 350W range even if the nameplate states otherwise. If you want to be conservative, use only 25% of the nameplate rating.

KEEPING THE RIGHT PERSPECTIVE

The purpose of this chapter is not to rail against the use of energy, but to minimize the environmental impact of energy consumption. When confronted with the disturbing effect of electrical energy on our environment, it is easy to assume this is a chapter against electricity use, the coal industry, or the electrical power industry. Quite the contrary, electricity is essential for business growth. Electricity creates efficiencies and helps businesses generate more revenue. Brave men and women risk their lives to mine the coal we take for granted when we turn on the lights. We should not feel guilty about using energy, or criticize the energy industries we rely on. We just need to increase our awareness of energy use and avoid needlessly using energy. If we don't manage our continually increasing consumption carefully, the increased consumption leads to a proportional increase of waste.

There is also some good news. The U.S. Electric Power Industry in conjunction with the Environmental Protection Agency (EPA) has greatly reduced emissions from power plants. For example, in 2003 the emission controls on the electric power industry resulted in a reduction in sulfur dioxide emissions of 38 percent from 1980 levels. New coal power plant technologies can reduce mercury emissions by about one half. And what is most impressive is that during 2003, the Office of Management and Budget found the United States Environmental Protection Agency (EPA) Acid Rain Program accounted for the largest quantified human health benefits, over $70 billion annually, of any U.S. federal regulatory program implemented in the last 10 years.[16]

SUMMARY

The needless consumption of electrical energy is not benign. As we have seen, power plant pollutants can be harmful, even deadly. The less energy we waste, the fewer harmful emissions are released into our environment. Reducing energy consumption may actually save lives. This motivates others to use less energy and reduces costs. We are more apt to curtail energy consumption when confronted with and reminded of the environmental aspects of energy consumption.

Some environmental effects are:

- In the United States, electrical power generation is the largest man made source of mercury in the environment.

- Electrical power generation is by far the largest single source of sulfur dioxide emissions in the US.

- Electric power plants in the U.S. create more carbon dioxide than the transportation sector. Carbon dioxide is a major cause of global warming.

- Gaseous emissions from electrical power plants may be the annual cause of:
  - — 30,100 deaths.
  - — 20,100 hospitalizations for respiratory and cardiovascular causes.
  - — 7,000 asthma-related emergency room visits.
  - — 18,600 cases of chronic bronchitis.
  - — 600,000 asthma attacks.
  - — 5 million lost workdays.
  - — 26 million minor restricted activity days.

- Energy reductions decrease emissions from electrical power plants and subsequently reduce the adverse health effects caused by electrical consumption.

References

1. Holiday Inn Hotels, http://www.4sight.org.uk, (February 18, 2006)
2. The World Bank, *2005 The Little Green Data Book*
3. U.S Environmental Agency, March, 2005. More information at www.epa.gov/mercury.
4. U.S. Environmental Protection Agency, "Acid Rain Program, 2003 progress Report," September 2004, Page 3.
5. U.S. EPA
6. U.S. EPA
7. "Energy Information Administration/Annual Energy Review 2004," www.eia.doe.gov. Table 12.2 Carbon Dioxide Emissions From Energy Consumption by Sector, 1980-2003,
8. Carbon Dioxide Information Analysis Center, Atmospheric carbon dioxide record from Mauna Loa, http://cdiac.ornl.gov/trends/co2/sio-mlo.htm, (June 22, 2006)
9. ibid
10. Source for the number of passenger cars (per 1,000 people), The World Bank, *The Little Green Data Book 2005*.
11. World Business Council for Sustainable Development, "Pathways to 2050; Energy and climate change," November 2005, Publications are available at: www.wbcsd.org and www.earthprint.com (June 22, 2006)
12. Full report prepared by Abt Associates, Inc. with ICF Consulting, and E.H. Pechan Associates, Inc., "The Particulate-Related Health Benefits of Reducing Power Plant Emissions," prepared for Clean Air Task Force, October 2000. A summary is available from the American Lung Association, Death, Disease, & Dirty power, September, 2000, http://cleanairstandards.org/artical/articleview/18/1/27/, (June 18, 2006)
13. U.S. Energy Information Administration, "Updated State-and Regional-level Greenhouse Gas Emission Factors for Electricity, 2002," http://www.eia.doe.gov/oiaf/1605/e-factor.html (June 22, 2006)
14. http://www.epa.gov/methane/
15. http://www.epa.gov/methane/pdfs/nitrousox.pdf.
16. U.S. Environmental Protection Agency, "Acid Rain Program, 2003 progress Report," September 2004.

Chapter 8

Critical Success Factors Towards Heightened Energy Awareness

Although technological changes in equipment can help to reduce energy use, changes in staff behavior and attitude may have a greater impact. Staff should be trained in both skills and the company's general approach to energy efficiency in their day-to-day practices. Personnel at all levels should be aware of energy use and objectives for energy efficiency improvement. ...Most importantly, companies need to institute strong energy management programs that oversee energy efficiency improvement across the corporation.

—Christina Galitsky, Ernst Worrell and Michael Ruth
Environmental Energy Technologies Division at
Lawrence Berkeley National Laboratory, "LBNL-52307"

Numerous studies show that the European Union could save at least 20% of its present energy consumption, or 60 billion euros a year through cost-effective energy saving measures.[1] The potential energy savings in the United States is probably much greater.[2] But saving energy takes a conscious and concerted effort. Armed with the knowledge from prior chapters, we are now well prepared to make that effort and reduce energy costs across our entire organization. This chapter is for the person that desires to participate in a formal energy awareness initiative. It is time to apply our newfound knowledge using proven techniques we will call "critical success factors."

You may be happy be learn that the critical success factors are not a one-size-fits-all process, or even a process, but rather essential components that should integrate into your existing organizational processes. You do not need to adapt your organization to a specific energy awareness process. The awareness success factors should be adapted to fit your organization's style and needs. An effective energy awareness initiative will leverage your existing processes and procedures where possible. Adapting the energy awareness program to your company's existing process and organizational strengths, will keep costs down and maximize the impact by leveraging existing infrastructure.

So, if your company uses Six Sigma techniques, incorporate the necessary energy awareness critical success factors into a Six Sigma system. Organizations that employ Six Sigma and total quality management techniques often find it easy to develop energy awareness initiatives. These organizations are accustomed to getting things done through guidance, empowerment and collective participation. A good example of this was published by the Alliance to Save Energy:

> *Through 2002, DuPont applied Six Sigma to behavioral tasks, including plant utility management. Over 75 energy improvement projects, many requiring no capital, were implemented across the company's global operations. The average project netted over $250,000 in annual savings*[3].

Again, whenever possible, try to leverage your organization's existing processes to spread energy awareness practices. Take a similar approach to funding. If your organization approves projects based on ROI (return on investment), pitch it that way; if your organization uses EVA (economic value add), Payback Period (the amount of time it takes for a capital project to recover its initial costs), or even Customer Satisfaction Metrics; fit the program into the current evaluation system. Since the programs are so inexpensive, a rigorous capital funding process is often not required.

Every organization should try to utilize all the basic energy awareness critical success factors outlined in this chapter. The actual implementation methodology of the critical success factors will vary in different industries. For example, universities have the challenge of reaching thousands of new students each year;

grocery store chains need to communicate with remote employees. This may necessitate a slightly different approach. Yet, a common dominator exists. All organizations are made up of people and the energy message needs to be conveyed to those people. Individuals use the energy at the university, the factory, the grocery store, and their behavior determines the efficiency of that use: A student could leave a dorm room computer on all night. A grocery employee could leave the walk-in freezer door open all night. A worker may fail to turn off air compressors at an idle factory. As we have seen, coercing people to change does not work, but the proper implementation of the critical success factors of an energy awareness program will.

The primary goal of the energy awareness critical success factors is to reduce energy expenses. This is accomplished by raising the organization's energy IQ and making everyone responsible for energy use. This small step is enormous. It signifies a paradigm shift to a new reality, that energy is a controllable cost. Companies that achieve six and seven figure energy cost savings do the following:

1. Make a serious commitment to saving energy.
2. Set energy reduction goals.
3. Communicate and take action.
4. Measure performance.
5. Repeat steps as necessary and stay ever vigilant.

CRITICAL SUCCESS FACTOR #1—EXECUTIVE LEVEL COMMITMENT

Programs that experience substantial and sustained energy cost reductions have executive level commitment. This is not unique to energy management. Quality and safety programs also consistently under-perform without strong executive backing. That's not to say that the actions of individuals can't make a difference. They can. Many company initiatives start in the trenches.

Senior level management can expand the energy saving process by:

1. Elevating the program credibility and importance.
2. Ensuring all members of the organization are aware of the initiative.
3. Assigning responsibility by disseminating energy goals and making all employees accountable.
4. Ensuring necessary budgeting and funding.
5. Guaranteeing employee suggestions are dealt with appropriately.

Designating an energy awareness leader validates that the campaign is a serious initiative and has executive support. The energy awareness leader should oversee the planning and implementation of the awareness program. Ideally, this employee will be a respected executive, or report directly to a senior executive. They should be interested in environmental issues. They should have the budget and authority to carry out their duties. The energy awareness leader may even be the CEO.

If employees think the energy awareness leader "is not committed" or "above saving energy" the program may not succeed. The leader must act as a good role model and exhibit good energy habits; people can sense insincerity and it will greatly inhibit success of the program. One tip for the awareness leader, invite employees to fine you $5 whenever they catch you leaving on your office light or computer monitor needlessly. You can give the employees the money, use it to fund an office pizza party, or donate the money to a charitable fund. Let them know you are backing up your commitment personally. Lead by example.

Of course, no single individual can perform all the functions required to run an effective energy awareness program. Leaders should assemble teams of committed people to manage the awareness program. It is important to decentralize the responsibility. The primary objective of the team is to influence people to use energy more responsibly, or simply: increase employee awareness of energy consumption and reduce energy costs without adversely impacting comfort or quality.

The typical size of these teams can vary. Membership can be permanent or temporary. Try to staff the team with individuals from many disciplines. Utilizing cross-functional teams ensures a comprehensive understanding of the various issues in all parts of the organization and facilitates fact-finding. And, equally important, involving and communicating with as many people as possible encourages buy-in and active participation once the program is underway.

Consider adding representatives from all levels of management. Include the following disciplines where appropriate:

1. Energy Management
2. Engineering
3. Finance or accounting
4. Front line staff such as:
   a. Assembly line workers and equipment operators in manufacturing
   b. Customer service in retail environments
   c. Housekeeping; bell staff and doorman in hotels
   d. Nurses and volunteers in hospitals
   e. Servers in restaurants
   f. Teachers and students in educational institutions
5. Human Resources
6. Information Systems (IS) or Information Technology (IT) departments
7. Maintenance
8. Marketing; Advertising; Public Relations; or Public Affairs
9. New Construction Department
10. Operations
11. Outside contractors who work in the facility
12. Purchasing or Supply Department
13. Sales
14. Security or Campus/Military police
15. Suppliers and Vendors

16. Utility Company or Energy Supplier.

Once the team is formed, it usually seeks a name or identity for the campaign. A name makes the program "real" and it facilitates communications, but don't spend a lot of time, or money, on names or logos. It can be as simple as the energy awareness initiative, or a bit more creative. Sample energy awareness campaign monikers include:

1. Energy Champions—Verizon
2. Flex Your Power—State of California
3. KILL-A-WATT—Iceland Frozen Foods
4. Lifeblood—Economic Development & Transport Minister, Wales
5. Operation Energy—Fort Lewis
6. Power Down—Bank of America
7. Reduce the Juice—A student lead initiative to reduce the electricity consumption of the town of Shelburne by 5%.
8. Shut It Off—United Technologies Corporation
9. Summer Survivor—Unisys
10. You have the Power—U.S. Department of Energy
11. Helping the Earth—British Telecommunications PLC

It's nice, but not essential to plaster the campaign name or logo on all the communication collateral. For example, high quality off-the-shelf posters denote a level of professionalism and commitment. It is necessary to replace these posters periodically to keep the program fresh; if they must be custom printed, it increases labor and costs. Try to keep things simple, the more steps in the process, the more opportunities to procrastinate, or delay getting the needed material. If you really feel the need, for consistent branding, to label every component of the program, try buying stickers with the logo and program name; and place the stickers on collateral. Often there is a desirable spot on the posters reserved for the stickers. This provides the best of both worlds.

The steering group should be tasked with:

1. Training for the steering group members.

2. Developing relevant measurement metrics and reporting system for announcing energy saving progress and team actions. This may involve recruiting other personnel or "energy champions" to assist with the program.

3. Raising the organizational energy IQ by dispelling myths and informing employees about energy consumption and their company's commitment to energy efficiency and environmental concerns.

4. Responding to suggestion box input and general questions. Always invite suggestions and feedback. Never criticize any suggestion or idea, keep the initiative positive.

5. Continually evaluating and seeking out new energy reduction and training opportunities. Initial simple low cost actions include: turning off unneeded equipment, proper maintenance and configuration of existing equipment, and ensuring that future equipment purchases are energy efficient.

Identifying opportunities is not as hard as it sounds. Once the program has begun, employees or other members of the organization will find and present many untapped saving opportunities to the team. Remember the Rover Group example in Chapter 6? That's the company that saved well over a million dollars. How? In part, by acting on a large number of energy suggestions. The number of energy suggestions received the month following the program launch exceeded the number received annually before the start of the program. So, don't try to solve all the problems up front. It is not the responsibility of the energy team to vet every opportunity before launching the program.

The team can start with a standard campaign of promoting

the importance of saving energy and the impact people have on energy consumption. Pick a small handful of ideas, such as listed in Section II of this book, to get people started in the right direction. In organizations with lots of personal computers, target awareness of computer energy consumption. Where occupants have control over space lighting, stress the impact of turning off the lights. Keep the number of targeted behaviors small, fewer than five. Do not try to do too much at once. Let everyone know tracking and monitoring of energy usage has begun.

If possible, consider having an energy audit. The audit may identify additional energy saving opportunities. Most audits include a quick check for lights and equipment left running when not needed. Ask the auditor to benchmark and compare your organization's energy uses against other similar organizations. Consider asking the auditor to ascertain the organization's attitudes and knowledge of energy efficiency.

If desired, the energy team can survey the existing level of energy awareness within the organization to potentially identify additional saving opportunities. Questionnaires are especially prevalent in large functionally diverse organizations such as universities. The education sector has a high turnover of students each year and each new class may exhibit different challenges. Schools may also have the advantage of using volunteers to design, issue, and analyze the questionnaires, sometimes for class credit. There is no productivity loss to schools when students take the time to answer questions.

If you decide to use a questionnaire, this is the start of the program. Start measuring savings from that point on, just asking questions increases motivation and reduces the energy bills. If done properly, the questionnaire can also communicate useful information and start the education motivational process. The advantages to starting the program with a survey are:

- A kickoff survey provides the opportunity to compare both "before and after" questionnaires to gauge the effectiveness of the campaign and make the next round more effective.

- It raises the level of employee ownership in the program, especially if used to select the program name or logo. It may also be more cost effective than forming focus groups to gather this level of involvement.

- The feedback can be a valuable aid in selecting both the target audience and appropriate messaging.

Some words of caution. Surveys require time, funding, and expertise to develop and analyze. People often don't want to respond to surveys. The information gleaned from questionnaires can be extraordinarily inaccurate. Energy managers perform physical audits of facilities and take careful measurements. Auditors do not ask if equipment is turned off. They check and verify that it is off. Never use survey results to gauge energy conservation activities as a high percentage of false answers are provided. Questions about forward-looking attitudes and intentions are also notoriously inaccurate indicators of actual behavior. Questions like, *"Do you plan to start turning off lights?"* are not useful. Only use the pre-program questionnaire to determine levels of energy knowledge or solicit input about the direction of the program. If desired, follow-up questionnaires can be used to gauge the impact of the program on energy savings, attitudes and motivation. Ultimately, the real test of the program is documented by a lower utility bill.

If you decide to take a survey, start with a simple and short questionnaire. Consider asking questions of the following nature:

ENERGY SURVEY

1. The United States is about four percent of the world's population. What percentage of the world's total annual energy usage does the United States consume?
   a) 5 percent.
   b) 10 percent.

c) 15 percent.
d) 25 percent.

2. Which fuel is used to generate the most energy in the United States?
   a) Coal.
   b) Oil.
   c) Natural Gas.
   d) Nuclear.

3. Which fuel is used to generate the most electrical energy in the United States?
   a) Fossil fuels such as coal, oil, and wood.
   b) Nuclear power.
   c) Solar and wind energy.
   d) Hydro electric power plants.

4. Scientists say the fastest and most cost-effective way to address our energy needs is to:[4]
   a) Develop all possible domestic sources of oil and gas.
   b) Build nuclear power plants.
   c) Develop more hydroelectric power plants.
   d) Promote energy conservation.

5. Which places more carbon dioxide emissions in the air?
   a) The electrical power industry.
   b) The transportation sector.
   c) The manufacturing sector.
   d) The computer industry.

6. The largest human caused source of mercury emissions in the air and sulfur dioxide emissions in the United States is:
   a) The electrical power industry.
   b) The transportation sector.
   c) The manufacturing sector.
   d) The computer industry.

7. On average, leaving on one computer every night and on weekends places the following emissions in the atmosphere:
   a) 1/3 pound sulfur dioxide and 2 pounds of carbon dioxide.
   b) 1/2 pound sulfur dioxide and 5 pounds of carbon dioxide.
   c) 1 pound sulfur dioxide and 1,000 pounds of carbon dioxide.
   d) 3 pounds sulfur dioxide and 2,000 pounds of carbon dioxide.

8. How much money can be saved per year by turning off one computer and monitor every night and on weekends?
   a) $5 or less.
   b) $25.
   c) $50.
   d) $75 or more.

9. Turning off florescent lights when leaving a room is:
   a) Not a good idea since turning lights off shortens their life.
   b) Not a good idea as turning lights back on uses more energy than leaving them on.
   c) Answers a and b.
   d) A great idea to save energy.

10. When a computer is turned off:
    a) The computer itself is still partially powered and using energy.
    b) The monitor attached to it continues to use energy.
    c) The speakers and other devices connected to it continue to use energy.
    d) All of the above.

11. Transformers and power supplies, those small blocks attached to the power cord of many appliances and electronic devices:

a) Do not use energy when the appliance is turned off.
b) Continue to consume energy even when the appliance is turned off.
c) It depends on the appliance. The transformer or power supply may, or may not, use energy when the appliance is turned off.
d) Use so little energy that it really doesn't matter.

12. Organizations that truly maximize energy savings:
    a) Utilize technology.
    b) Solicit help of employees.
    c) Have high organizational energy IQs.
    d) All of the above.

13. We are launching an energy awareness campaign; please pick your favorite campaign name:
    a) You can make a difference.
    b) Operation save the day.
    c) Turn it off.
    d) Energy Awareness Campaign.
    e) Other __________

14. What would be a good incentive to encourage employees to save energy?
    a) Prizes.
    b) Awards.
    c) Contests.
    d) More awareness, education, and reminders.
    e) Don't know.

15. What percentage of time do office machines idle in a normal workday?
    a) Up to 15%
    b) Up to 45%
    c) Up to 65%
    d) Up to 90%

16. How much do you think this organization spends on energy annually?
    a) $500,000.
    b) $1 million.
    c) $10 million.
    d) Other _________

Correct Energy Survey Answers

1. D. The United States is responsible for 25 percent of the world's total annual energy consumption.
2. B. Oil generates most of the energy in the United States.
3. A. Burning coal creates most of the electrical energy in the United States.
4. D. The fastest most cost-effective way to address our energy needs is to promote energy conservation and use energy wisely.
5. A. The electric power industry places more carbon dioxide in the air than any other sector.
6. A. The electric power industry places more mercury and sulfur in the air than any other sector or industry.
7. D. Needlessly leaving on one computer on nights and weekends can place 3 pounds of sulfur dioxide and 2,000 pounds of carbon dioxide in the air.
8. D. On average, leaving that computer on costs your company $75 per year. If your company has 5,000 personal computers, that's $375,000 a year. This does not include the computer monitor, speakers, or other devices connected to it.
9. D. It's a good idea to turn off the lights when you leave a room.
10. C. Computers use energy even when turned off. Some of this is the power supply; some is standby power on the motherboard. The devices connected to the computer also continue to consume energy.
11. B. Transformers and power supplies use energy whether or not the device they are powering is energized. We spend several billion dollars each year powering "off" devices.

12. D. All of the above. Organizations that truly maximize savings solicit help from employees and have high organizational energy IQs.
13. The answer is subjective and will be unique to your organization.
14. The answer is subjective and will be unique to your organization.
15. Office machines idle up to 90% of the time.
16. The correct answer will be unique to your organization.

Scoring

Give one point for each right answer to get the final energy IQ score.

13-16: Great. This should be a typical energy manager score.
9-12: Good. Good knowledge about energy consumption.
0-8: No worries. Just expedite the energy awareness program.

Prompt follow-up is also desired. If a substantial amount of time passes between the survey and campaign launch, members of the organization will conclude it is not a serious initiative.

CRITICAL SUCCESS FACTOR #2— SET ENERGY REDUCTION GOALS

If you don't know where you are going...
You might end up someplace else...
—Yogi Berra

Making energy awareness an integral part of business planning often yields the greatest benefits. It may be appropriate to add the energy efficiency targets as one of the corporate-wide key performance indicators (KPI), or include energy efficiency results in the annual performance reviews of key employees. Ultimately, your organization will need to find its own best approach.

Please fight the urge to start with large energy saving goals that may require coercion or employee stress. Energy awareness initiatives are inexpensive and we can afford to start slow and build a revenue positive lasting campaign. It is generally a good idea to start with an easy energy cost reduction target of perhaps just 1 to 3% of overall energy use and achieve quick success. Nothing breeds success like success. Also, consider expressing energy costs as a percentage of controllable cost versus total costs. In some manufacturing facilities, energy may only represent 3 to 5% of cost, but up to 40% of controllable cost.

Make financial goals personal, relevant, and easy to understand. People will take more ownership of goals, if someone else does not unrealistically impose them. This is counter intuitive, but I've found, through personal experience, that employees will often set higher goals for themselves than managers would have imposed. Often, bilateral goal setting works best. The energy team should try to include as many people as possible in the goal setting process. Use the documented savings examples of other similar organizations to inspire participation. Remember, this is a continual process. It is better to start with a solid foundation and gain momentum, as the right habits are formed and people see success, the initiative will grow.

The specific KPIs help measure progress towards organizational goals, in this case, reduced energy expenses. Most employees will be surprised to learn the cost of energy and the amount of pollution they generate. Try to incorporate actual energy costs and emissions in the KPIs, rather than just energy or kilowatt reduction goals. Consider using KPIs as shown below:

Potential Goals

- Energy, energy costs, or emissions per unit of production such as CO_2/widget, kWh/widget, Btu/assembly line, $/ton output, or CO_2e/gallon.

- Energy, energy costs, or emissions per square foot (or meter) such as CO_2/square foot or, $/Sqft, and Btu/Sqft/year.

- Energy, energy costs, and emissions per hotel room such as CO_2/hotel room or $/hotel room, or kBtu/hotel room/year.

- Energy costs and emissions per student such as CO_2e/student, $/student, kBtu/student/year.

- Energy costs per food count (meal served) such as CO_2e/food count or $/food count.

- Comparison to similar buildings and manufacturing plants (ranking or percentile).

- Percent to revenue goal. Reduce energy costs by 10% per dollar of revenue between 2008 and 2010.

- Percent improvement verses prior year, such as reduce energy by 25 percent below 2004 levels by 2010.

- Energy usage to a fixed number such as energy costs will not exceed $56.6M in 2010.

Don't create goals that can't be tracked. Setting and following metrics is critical. Failure to report results or inaccurately reporting results can negatively affect morale and induce the wrong behavior. Try to determine what can be tracked and use those Performance Measures (PM) instead. They may not be perfect, but it's a start. And consider adding non-financial goals for the energy team. Some potential metrics are: the overall increase in the organizational Energy IQ, time to respond to employee suggestions, and the number of low cost initiatives implemented. In the future, continually roll some of the energy cost savings into developing a better reporting and measurement system.

CRITICAL SUCCESS FACTOR #3— COMMUNICATE AND TAKE ACTION

The overall success of the energy efficiency initiative depends on the cooperation, acceptance, and involvement of as many people as possible. In many organizations, this extends beyond direct employees to students, patients, visitors, customers, guests or tenants. It is important to identity the target audiences, and then, choose the best communication tools to reach the desired audience. Once the audience sub-segments are delineated, choose the appropriate message and the optimal communication delivery method. This will vary by organization type. I will list a few examples to get you started:

1. Tenant Occupied Office Space

If you are a "tenant occupied" property owner, the cleaning and security staffs working for you are one audience. You will encourage them to turn off lights after hours while making their rounds and track results though the whole building main meter. Another audience may be the tenants: You may decide to target their behavior through posters and brochures encouraging them to turn off devices and report energy waste through an 800-phone number. You may even offer to let them sub-meter their own office space. You may choose to provide a real-time energy usage data via a web page. Remember, well run energy awareness programs do not disturb tenant comfort and usually increase tenant satisfaction. Communicating an energy awareness campaign may look like Figure 8.1.

2. University

In a university, you may decide to use the target audiences' communication methods, Figure 8.2, to deliver a turn off the lights when leaving a room message.

The importance of communications should not be underestimated. Effective communications is essential to the success of the project. A successful campaign will increase awareness, interest, and desire, which leads to action and energy savings. Some methods at your disposal may include:

The Personal Touch

Small Face-to-face Meetings

Meetings are one of the most effective forms of energy awareness communication. For normal campaign updates and

Audience Segmentation	Communication Method							
	Direct Letter	Fine Art Poster	Bulletin Board Poster	800 Phone Number to Report Waste	Suggestion Box	Meeting	Track Results with Sub-Meter	Track Results with Master Meter
Tenant	X	X		X		X	X	
Guests		X			X			
Security			X		X	X		X
House Keeping			X		X	X		X
Maintenance			X		X	X		X

Figure 8-1

reinforcement, consider adding energy topics to other regularly scheduled meetings to reduce costs. Ideally, the energy program will become part of everyday operations and it will be natural to include energy awareness updates in general finance or business update meetings.

Dedicated Presentations

Consider kicking off your awareness program with a special event to get the organization fired-up: invite employees to a free luncheon. Have a senior executive say a few words to emphasize the corporate sponsorship, and invite a professional energy awareness expert to speak. The presentation should outline the reason for the program and potential results. Remember, the saving potential for even medium size facilities is in the hundreds of thousands of dollars per year. This type of fanfare builds employee good will and morale.

Internal Training

A training class covering all the topics discussed in the book may be necessary. Individuals need to have a comprehensive understanding of the cost and impact of energy consumption, and the barriers that allow energy waste to flourish. The more knowledgeable they are, the easier it is for them to pass the knowledge on, overcome objections, and sell the campaign to others. The training will also aid in the implementation of specific actions.

Informal Gatherings

Energy champions should take advantage of ad-hoc meetings

Audience Segmentation	Communication Method						
	Campus Newspaper Articles	Fine Art Posters	Bulletin Board Poster	Contest	Workshops	Promotional Gifts	One-On-One Discussions with Volunteers
Students - Postgraduate	X	X	X				
Students - Undergraduate	X	X					
Student - Part-time	X	X					
Students - Full-time	X	X					
Students - Resident	X	X		X		X	X
Students - Non-resident	X	X					
Staff - Academic	X	X			X	X	
Staff - Nonacademic	X	X			X	X	
Administration	X	X				X	
Contractor	X		X		X		
Staff - Technical	X		X		X		
Housekeeping	X		X		X	X	
Security	X		X		X	X	

Figure 8-2

to solicit feedback on the campaign progress or solicit energy saving ideas.

Door-to-door Canvassing

Universities often use student volunteers to actively seek out and discuss saving opportunities with dorm residents.

Competitions

Consider friendly competitions between different parts of the organization, or awarding small prizes for the best results.

Suggestion Box

Provide a method for individuals to share their cost saving ideas. This can be an actual physical suggestion box, a telephone hotline, an e-mail address set-up especially for this purpose, a blog, or an interactive web site. Always acknowledge and provide a thank note for each suggestion. Always stay positive. There is never such a thing as a stupid energy question or idea.

Videos, Web Casts, and DVDs

The cost to create custom videos and DVDs has dropped dramatically. Consider filming parts of the dedicated presentations or training classes. This ensures viewing by a wider audience.

Print Material

Direct Letters, Newsletters, Pamphlets and Brochures

Pamphlets or brochures are good at launching campaigns and for use during new employee orientation. Often they are short, three fold 8-1/2 x 11 sheet, or they can be quite lengthy. In a building owner and tenant situation, announcement letters to tenants may be sufficient.

Newsletters are a low cost approach to offering new ideas, reinforcing ongoing messages, and reporting progress. Basic newsletters can be used at any stage of the campaign and may be distributed via e-mail to reduce costs. They should include basic tips on saving energy, costs of operating common equipment at the facility, overall energy cost, environmental information, and the companies goals and objectives. In later stages, include scorecard information such as performance to goal, or comparisons to other sites for friendly competition. Section II provides weblinks to example material, some of which you can download and modify.

Books

This book was designed to be read quickly, by all level of employees, and yet be comprehensive enough to provide a thorough understanding of the issues. Reference material, such as motivational quotes and energy tips, is separated for later reference as time allows keeping the motivation level high and growing the program. Use this as a tool for energy management champions and personnel resistant to change. For these people the investment in additional information is warranted.

External Publicity

It is sometimes useful to issue press releases to the local media about the program and its progress. Be imaginative and name individuals that made special contributions. Some organizations also sponsor publicity events such as fun runs or walks. Or leverage public events such as Earth Day by having parties or picnics. Interest from staff, suppliers and families will keep the momentum going.

Posters

Professional posters are phenomenal at attracting attention and raising awareness. They can be used at any stage of the campaign and even used as a stand-alone campaign. Place them in high traffic area such as cafeterias, lobbies, elevators, waiting rooms, hallways, and by vending machines. Posters should present simple and direct messages.

There are typically two kinds of energy awareness posters:

1. Small 8 x 11 or 11 x 17 inch posters designed for work areas and bulletin boards. Some of these can be downloaded free off the Internet and printed in your office. They should be changed

at least four times a year to keep the messages fresh. Section II provides web-links of example material, some of which you can download and modify.

2. Fine Art poster. Larger and framed, fine art posters attract attention, and drive results letting both the organization and visitors know the cause is serious. They are suitable for offices and common areas. Posters can become invisible over time. They should be changed several times a year to keep the messages fresh.

Stickers

Small stickers are useful on light switches and some computers. They can become invisible or ignored, if not changed from time to time. Consider only buying easy to peel off stickers (Post-It® notes) and replacing them from time to time. In areas with high summer demand periods it is useful to put them on at the beginning of summer, then peel them off at the end of summer, repeating the process every year with new and relevant stickers can result in energy savings.

Websites

Consider creating an internal energy management website. In addition to displaying a wealth of information, websites can display up to date information on energy usage and billing data.

CRITICAL SUCCESS FACTOR #4 MEASURE PERFORMANCE

In an aluminum foundry, the production director placed a board outside the canteen door and wrote on it: "Last week we used 74 gallons of oil per tonne of aluminum. Our target is 49." Within three months the target had been achieved, a savings of 34%. And further savings are still being made.

—Carbon Trust[5]

Measurement is essential. If we can measure it, we can manage it; if we can manage it, we can improve it. Monitoring efficiency projects and behavioral energy habits is a necessary step. Without monitoring, savings will erode over time. Continual measurement of the awareness initiative helps us:

- Quantify the return on investment
- Gauge awareness and compliance
- Document successes
- Identify opportunities
- Improve existing processes and deliver more value
- Spot when mid course corrections are necessary
- Reward good performance
- Maintain program momentum

Every program will have its problems. These are difficult to predict and may be exceedingly minor. Always point out the successes when presenting the data. A vital pillar of leadership is the ability to gather, assess and understand the right data to effectively drive change. Regularly provide the program results to the entire organization. Nothing breeds success like success.

FACTORS THAT AFFECT ENERGY CONSUMPTION RESULTS

Several factors can skew the awareness campaign results. We need to acknowledge and account for events that can alter our results. Often, what we are measuring is avoided cost, rather than actual costs.

Changes in energy tariffs

If you are tracking results in dollars, be sure to factor any tariff charges into the savings calculations. For example, if an energy initiative reduces energy consumption by 20%, but energy costs rise 20%, the bill will be the same. Yet, had the initiative not occurred the energy bill would be much higher.

Energy Upgrades

Be sure to check for any energy retrofit projects such as lighting or heating ventilation and air conditioning upgrades that may also reduce energy usage. Try to delineate the effect of each energy saving initiative. This may require the addition of metering equipment or an analysis by an energy manager.

Occupancy

Occupancy, or the number or persons in a building can greatly impact energy consumption. Be sure to factor population swings into the savings calculation. This also includes one-time special events and occasional large gatherings.

Production

Similar to occupancy, the amount of product a factory produces tends to increase energy costs. Energy awareness programs are often effective enough to reflect an absolute savings despite increased production. Still, it is better to arithmetically eliminate the effects of the increased production.

Weather

Outside temperature changes can have a large impact on energy bills in commercial and residential buildings. When measuring the awareness campaign results, differences in the weather from our baseline year and our campaign period can make the outcome appear artificially high or low. If you suspect there are temperature differences between a given month this year and last year, there are software packages that use statistical analysis to remove the effects of weather and provide good performance comparisons. Providing weather corrected energy reporting is an important function or temperature-related effects can distort the savings efforts.

Maintenance

Neglecting routine maintenance, will of course, increase the amount of deferred maintenance. Increased mechanical inefficiencies can negate the savings of employee's efforts turning off devices. If you suspect this is an issue, be sure to track and look for correlations between deferred maintenance and energy consumption.

Slowly Changing Dimensions

Often small changes can creep up on us. For example, we tend to think of square footage as a constant value. Nevertheless, over time it can increase slowly, especially at universities, which always seem to be adding buildings. Be sure to track and save such slowly changing dimension. A university calculating the energy per square foot should use the square footage that existed at the same point in time as the energy measurement occurred. Sounds like common sense, but it is easy to neglect, especially in dynamic environments with lots of little changes.

CRITICAL SUCCESS FACTOR #5 REPEAT STEPS #1 THROUGH #4 AND STAY EVER VIGILANT

New products, new technologies, employee turn-over, and organizational changes will both help and hinder the energy awareness initiative. Be sure to continually monitor the energy savings and communicate the results to the organization. It is important to stay vigilant and respond to problems and opportunities. There are also seasonal considerations. Be sure to plan for the summer peak demand periods, leverage Earth Day events, or other regularly scheduled meetings. Posters and communication collateral should be updated regularly or they effectively become invisible.

Consider periodic changes to the measurement goals and metrics. This will keep the program fresh and relevant. If the initial program energy goal was "percent energy savings verses prior year energy consumption" a change to "dollars of energy consumption per square foot" might be a good replacement metric for the next round of savings. For example, asking a student dormitory to reduce greenhouse gas emissions by ten percent over the same period in a prior year is a good starting point. After running the program for some time, the student dormitory may achieve an excellent energy record. At this point, asking the next residents to improve on this by another ten percent may be ineffective and set the class up for failure, possibly increasing energy usage. However, asking the new class to maintain the prior benchmark in greenhouse gases per dorm room and rewarding that behavior may be advantageous. In any case, occasionally changing the metrics maintains participant interest.

SUMMARY

Energy awareness initiatives work best when they are incorporated into existing organizational processes. It is essential to include the following critical success factors in to any existing process methodology:

1. Make a strong commitment to saving energy.
2. Set energy reduction goals.
3. Clearly communicate expectations and take action.
4. Measure results and reward or acknowledge performance.
5. Repeat steps as necessary and stay ever vigilant.

Done properly, an energy awareness program will cause the organization to focus on energy efficiency and reduce energy costs in a sustainable manner.

Bibliography

U.S. Environmental Protection Agency, "Guidelines for Energy Management." It recommends a process as follows: Make a commitment → Assess performance & set goals → Create action plan → Implement action plan → Evaluate progress → Recognize achievements. Continually re-assess progress and continue the process. More details are available from Energy Star in an Adobe Reader PDF format at www.Energystar.gov

Natural Resources Canada, "Saving Money Through Energy Efficiency, A Guide to Implementing and Energy Efficiency Awareness Programs." It recommends a process as follows: Assemble the players → Identify awareness program opportunities → Establish objectives of the energy efficiency awareness program → Develop a communications plan → Implement your energy efficiency awareness program → evaluate your program → track and report results → follow through. More details are available at the Natural Resources Canada Web site http://www.nrcan-rncan.gc.ca.

U.S Department of Energy (DOE) Federal Energy Management Program (FEMP), "Creating an Energy Awareness Program," A Handbook for Federal Energy Managers. It recommends a process as follows: Plan the Effort → Design and Implement the Program → Evaluate and Report Results → Sustain the Effort. More details are available at the U.S. Department of Energy Federal Energy Management Program http://www.eere.energy.gov/femp/.

Carbon Trust, Carbon Management, "Assessing and managing business responses to climate change by Carbon," June 2005. Carbon Trust highlights a high Level Process used in conjunction with UK companies as follows: Mobilize the organization → Evaluate the business case → Identify opportunities → Develop implementation plans → Manage implementation. More information is available at http://www.thecarbontrust.co.uk/carbontrust.

State of California, Flex Your Power Program, "Business Guide 3: target Business Employees for Energy Conservation in the Workplace," N.D. Flex Your Power recommends the following high level process: Devise a plan → Implement programs/operations → monitor and measure results

Power Smart awareness program at BC Hydro website. BC Hydro recommends a six step program as follows: Step 1—Establish a clear vision → Step 2—Create the team → Step 3—Know your workplace → Step 4—Develop your communications plan → Step 5—Implement your plan, and Step 6—Recognize and reward

References

1. European Commission, "Doing more with less, Green paper on energy efficiency, 2005"
2. Ibid; the energy intensity (tons oil equivalent/million euro of GDP) indicates more energy efficiency in the EU than the USA.
3. Christopher Russell, "ENERGY MANAGEMENT PATHFINDING: Understanding

Manufactures' Ability and Desire to Implement Energy Efficiency," March 3, 2005, P. 3.

4. National Environmental Education & Training Foundation (NEETF), "Americas' "Low Energy IQ:" A Risk to Our Energy Future: Why America Needs a Refresher Course on Energy: The Tenth Annual National Report Card: Energy Knowledge, Attitudes, and Behavior," August 2002, P. 16.
5. A booklet by the U.K Department of the Environment, Transportation and the Regions, "Managing and Motivating Staff to Save Energy, Good Practice Guide 84," March 1999, P. 11.

Chapter 9

Section Wrap-up

Through elementary energy efficiency steps DuPont, the chemicals group, has saved up to $2bn since 1990.

—Fiona Harvey, *Financial Times*

We began with an everyday observation that many individuals habitually fail to turn off unneeded lights and computers. This observation raised a simple question: Should we manage this behavior, or merely ignore it? Financially, leveraging the entire organization to reduce energy costs is a worthwhile endeavor. Companies that manage the behavioral aspects of energy consumption often reduce energy costs 15 percent or more through simple low-cost, or no-cost, non-technical energy saving measures. Payback periods are short, often expressed in only weeks or months. As energy costs rise, the decision to manage the behavioral aspects of energy consumption becomes increasingly attractive.

Integrating behavioral and non-technical energy saving activities into "core" planning and management processes is a straightforward task. Begin by explaining (or learning) the basic reasons why electrical energy is so easy to waste, take the time to manage attitudes and mind-sets, and eliminate any misinformation or myths. Utilize the included financial examples and quotes to show the tremendous savings potential and garner the support of the organization. Be sure to leverage the environmental consequences of energy consumption. Even if environmental concerns do not motivate you, they do motivate other people to reduce energy consumption. This is especially helpful at universities, hotels, tenant occupied offices, and other locations where the users of energy are not direct employees.

Try to integrate the critical success factors in Chapter 8 into your existing processes and procedures, starting with a strong commitment from management. One of the newer trends in business is renewed emphasis on measurement and quantifying. This aligns well with energy awareness initiatives. Be sure to set energy consumption goals and document the organizations performance. Energy managers have long understood the importance of tracking energy costs as a feedback tool for motivating and harnessing the entire organization. Finally, continually communicate the importance of reducing energy costs across the entire organization.

Section II

Reference Section

Chapter 10

Low Cost and No Cost Actions

Adobe's simplest—and by far the cheapest—energy-saving solution was to reduce the operating hours of garage exhaust fans and outdoor lighting systems. The cost: $150; annual savings: $68,000.

Jeff Nachtigal
"The Greenest Office in America"

People and ideas are the driving force behind energy efficiency. Try to mobilize the entire organization to identify energy saving opportunities. Look for quick easy wins. The examples in this chapter are standard tips to get you started and thinking in the right direction. Choose half a dozen items to start with. Be sure to couple the specific assignments with high-level goals for energy reduction. Once people are accustomed to reducing energy costs and have goals to strive for, more powerful initiatives unique to each facility or manufacturing process can be explored.

OFFICES/OFFICE EQUIPMENT

Tenants directly influence over 50% of the energy consumed in a typical office building (including their impact on office equipment, lighting, and temperature). Office equipment uses about 3% of all the energy consumed in the United States.

No Cost Opportunities

- The computer monitor often uses more energy than the computer. Turn off your computer-monitor (screen) when not in use, even if it's just for 15 minutes. The monitor has a quicker restart time than the computer so it is more convenient to turn off during short time intervals. There is little downside to turning off a monitor no one is looking at.

- Consider turning off monitors and computers at the end of each day and utilize the sleep and hibernation features. Turning off computers on nights and weekends can save $80 per year in energy costs.

- Enable Energy Star features on your computer. Many computers are idle for long periods of time and sleep modes can save large amounts of energy. In 2003, Energy Star estimated that 45% of the computer monitors in the United States do not take advantage of built-in sleep features, costing businesses and organizations at least $225 million a year in wasted electricity.

- Close the lid on your laptop when you are not using the laptop. This can place the computer in a low energy state that wakes up quickly upon raising the lid. This has the added advantage of securing the computer, as the password typically needs to be reentered upon the power-up cycle.

- Install free software from the Environmental Protection Agency that puts monitors in sleep mode when not in use.

- Turn off photocopiers during off-hours. Operating photocopy equipment efficiently will reduce its energy use by 25% or more. Consider setting the copier to the energy-efficient setting, this will reduce the amount of energy consumed when it's not in operation. However, it is still using energy, so when practical, turn it off.

- To turn off computer scanners, printers and other devices that are plugged into a power strip, simply switch off the power strip after shutting off your computer.

- Unplug nonessential energy using devices such as coffee makers when leaving for the day.

- Consider banning space heaters, microwaves, refrigerators, and incandescent lights throughout the facility

- Do not leave equipment in sleep mode overnight because it will continue to draw a small amount of power. Turn off all equipment every night, especially monitors and printers. Monitors usually consume more electricity than CPUs.

- Use students, cleaning, or security personnel to turn off items such as office lights, copy machines, coffee pots.

- One university unplugged over 500 campus water coolers to save energy. Look for small nonessential loads.

Low Cost Opportunities

- Choose electronic products and appliances without a built-in clock or timer. The device's display only consumes about 1/2 watt, but the power supply and the appliance powering the clock may consume ten times as much energy.

- Use laptop computers. Laptops use up to 90% less energy than a desktop computer.

- Use flat panel screens. They use less electrical energy than standard CRTs. Some utilities offer rebates for the purchase of new ENERGY STAR® certified LCD monitors to reduce your energy costs. For example, Silicon Valley Power has offered companies rebates of up to $20 per monitor up to a maximum of $10,000 per customer. Check with your utility for current rebate programs.

- Print with ink jet printers instead of laser printers. Ink jet printers' cost less to maintain and use 90% less energy than laser printers.

- Replace existing equipment with Energy Star equipment.

LAMPS AND LIGHTING

Lighting typically represents about 30 percent of the total energy consumption in an office building.

No Cost Opportunities

- Train and encourage employees to turn off interior lights when they are not needed. The overwhelming majority of energy consumed by lamps turns into heat. Turning off the light will also keep a room cooler, and reduce the air conditioning bills.

- Remove lamps where you have more lighting than you really need, but be sure to maintain safe lighting conditions. Disconnect the unused ballasts.

- Harvest daylight and make use of the sun. If your lights can be controlled separately, consider turning them off whenever there is enough natural light available.

- Adjust lighting levels to match needs at different times. Full lighting may not be necessary when performing tasks such as cleaning.

- Make sure outdoor lighting is not used during daylight hours.

Low Cost

- Consider moving janitorial night cleaning to the day. One U.S. DOE presentation estimated that wasted lighting consumes approximately 24% of total lighting usage. The same presentation concluded that reducing janitorial services night lighting saves, on average, $14,400 annually per 100,000 square foot of office space. There is an optional small investment in tools, such as battery powered floor buffers so that there are no cords to trip over and silent vacuums so as not to disturb tenants.

- Meter and track daily energy consumption and costs. Understanding when and how energy is used in a building can lead to changes that reduce energy costs. For example, large night loads may indicate that lights and equipment are running unnecessarily. Simple benchmarks such as energy used per square foot (kBtu/sqft) can be compared to similar buildings or past usage.

- Install light switches where employees can easily access them.

- If large banks of lights are on a single switch consider installing more switches to improve control of individual zones

- Install automatic lighting controls. These devices help optimize lighting use by automatically turning lights, off, or dimming depending on occupancy, ambient light, or time of day.

- Consider the opportunity to install occupancy sensors to automatically turn off lights in conference rooms, closets, break rooms, demonstration rooms, and even restrooms.

- Buy energy efficient bulbs, fixtures, and ballasts. Convert T12s to T8s and T5s.

- Have security guards, cleaning staff, interns, or students check for unnecessary lighting.

- Eliminate magnet ballasts and go electronic.

- Replace inefficient "exit" signs with high efficiency LED "exit" lights. They use up to 95 percent less energy and last 10-20 times longer than incandescent bulbs.

- Dust and dirt on lamps and light fixtures cause a light loss called Luminaire Dirt Depreciation. Consider cleaning all surfaces and replacing old "yellowed" lens shields with newer acrylic lenses.

- When remodeling or refinishing consider painting with light colored or high reflectivity materials.

- Replace incandescent bulbs with compact fluorescent lamps. They use up to two-thirds less energy, last about ten times longer, and burn cooler lessening building cooling costs.

- Charge tenants for after hour lighting usage.

- Correct operation and maintenance (O&M) issues. Ensure that equipment is operating efficiently, thermostats are calibrated, dampers and adjustable frequency drives are properly adjusted. Effective O&M will increase occupants' comfort level and decrease equipment malfunctions.

HEATING VENTILATION AND AIR CONDITIONING (HVAC)

Heating and cooling costs account for approximately 40 percent of the energy costs in an office building.

No Cost

- The perfect temperature used to be when half the people in a room were to cold, and half were to hot. Now, typical winter thermostat settings may be 68°F or below during business hours and 63°F after business hours. Summer thermostat settings may be 72°F or more during business hours and 85°F or off after business hours in the summer. During the California energy crisis in 2001, businesses where asked to raise summer settings to 78°F when possible. Be sure to alter the settings of heating and cooling systems seasonally to save energy. Collaborate with tenants to find the temperature that is most comfortable and best meets other climate control requirements. You may find that some areas are warmer or cooler than they need to be.

- Keep exterior doors closed as much as possible. A large door left open can add several thousand dollars to the energy bill.

- Consider adjusting work schedules to reduce energy use during peak demand periods. In the summer start earlier or consider early morning precooling to a lower the temperature and allowing the temperature to slowly rise throughout the day to reduce air conditioning load during the hours of peak electricity demand.

- A good low cost way to reduce energy expenses is to negotiate a better utility charge. It will be easier to select another tariff once the organization has a better understanding of when, where, and how energy is being used

- Use shades and blinds to keep direct sun light out.

- Adjusting the fan speed can reduce energy use. A 10 percent reduction in fan speed can cut energy consumption by 27 percent.

- Make sure that furniture does not block airflow around radiators or air intakes and diffusers.

- Seal off unused areas. Don't heat or cool these areas more than necessary.

- After preheating the building in the winter (prior to occupancy), consider turning off the heat and allowing equipment and personnel (each person produces 500 to 600 Btu/hour) to heat the building. Large, well-occupied, commercial buildings seldom need much heat when the lighting is on, even in the winter.

- Adjust airflow so that portable electric heaters are not necessary.

- When remodeling or repainting buildings consider using lighter colors that reflect more sunlight, thus lowering air conditioning expenses.

- If tenants have not experienced energy awareness training, eliminate or minimize tenants' access to thermostats.

- Money can be saved by adjusting the preheat or precooling times. Collaborate with building occupants and experiment to determine the latest possible start-up time for HVAC equipment without affecting tenant comfort. In addition, determine the most efficient sequence of operation for the equipment.

- Verify that the energy management readings, from thermostats and other sensors, reflect what is the actually happening in the building.

- Make sure the economizers are functioning correctly. A properly functioning airside economizer uses outdoor air for ventilation when the outdoor air temperature is cool enough to replace or reduce energy intensive mechanical cooling. It's similar to opening a widow to cool a house. I have seen economizers permanently propped open. This causes excess air to be brought in. Alternately, if the economizer is permanently off, cheaper outdoor air cannot be used for cooling.

- Personal space heaters, open windows and doors, covered diffusers, and personal fans are an indicator of personal discomfort or equipment issues.

Low Cost

- Make sure that dead bands are correct so there is no simultaneous heating and cooling.

- Meter and track daily energy consumption and costs. Examine load profiles to identify potential problems or ways to reduce costs.

- Dirty air filters and cooling coils in an air conditioning system can reduce efficiency by 30 to 40% or more, keep them clean. A clean air filter has much less air resistance. The less air resistance there is the easier it is for the motors to move the air around buildings.

- Perform regular maintenance to keep heating, ventilation and air conditioning (HVAC) systems running more efficiently. Maintenance activities can save up to 30% of fan energy and up to 10% of space conditioning energy use. Check fan belts, lubricate motors, pumps and fans, check seals, clean condenser coils, and patch leaks. Verify adjustable frequency drives are configured correctly.

- Seal heating and cooling ductwork. Leakage is especially expensive if the air leak is in an area that does not require conditioned air.

- Seal exterior cracks and holes. Small cracks around doors and widows can really add up to large heating and cooling losses.

- Reduce air conditioning and heating hours by installing a time clock to turn off the system when the building is unoccupied.

- When fan and pump motors need repair replace them with energy efficient models.

- Relocate thermostats to optimal locations, such as by return air ducts.

- Install adjustable frequency drives.

- Consider running the cooling tower without fans. On mild days, the cooling towers can sometimes be used to provide natural cooling from convection without the fans.

- Add strip curtains to exterior openings and loading docks where necessary.

DATA CENTERS

Energy saving activity may seem off limits at data centers since reliability is so essential, however a survey by the Uptime Institute of 19 data centers found that 1.4 kilowatts of power are wasted for every 1 kilowatt consumed.

No Cost Opportunities

- Servers that support businesses must be kept at a constant and cool temperature to function properly. Thermostats in many data centers are set to 70°F or below, but setting the temperature to 74°F will reduce energy costs without affecting server reliability or performance. Check with the computer manufacture if you are unsure of the recommend levels. (From IBM e-magazine article.)

Low Cost Opportunities

- Consider consolidating the sever locations. Many small and medium size businesses have servers spread out in many locations in a building, or excess servers that require energy to cool and ventilate. Infrastructure consolidation can generate significant savings.

- Use Energy Star computers and utilize power features that can turn on and off computing power as required. No sense paying for unused resources.

- Alternatively, consider allowing an application service provider to host your applications. Outsourced applications are typically housed in dedicated data centers that manger power more effectively. In addition, you are protected against energy price spikes, as energy is not typically included in the hosting fees.

- Consult a qualified energy manager.

MANUFACTURING FACILITIES

No Cost Opportunities

- Encourage other employees to be energy conscious. Set-up a process for spotters to report steam, water, or compressed air leaks. Provide small rewards for spotters.

- Determine performance metrics and benchmark current activities using measurements such as amount of energy per finished product, amount of water and waste per unit produced. Continuously assess progress.

- Turn off unneeded equipment such as air compressors and assembly lines whenever possible.

- Compressed air is very expensive, in some cases it can cost as much as natural gas. Do not use it to sweep the floor. Manage the use of compressed air carefully.

- Reduce the pressure of the compressed air system to the lowest acceptable level.

- Manage loads to avoid setting new peak demand charges.

Low Cost Opportunities

- If possible, distribute the energy usage over many shifts, or move an energy intensive production to the second shift.

- Minimize HVAC equipment during peak demand periods.

- Ensure that the air compressor air intake is located outdoors or in the coolest location. The cooler the intake air the less energy the compressor uses.

- Discharge the air compressor cooling air outdoors in the summer and into areas requiring heat in the winter.

- Insulate steam lines.
- Analyze the use of adjustable speed drives on motors.
- Consider connecting energy consuming equipment directly to other process equipment so that the equipment only runs when the plant is in active production.
- Meter and sub-bill each process.
- Launch an aggressive program to fix steam and compressed air leaks. Never allow steam leaks to exist more than a few days.
- Reduce power factor charges through power factor correction capacitors.
- Evaluate an interruptible demand tariff.

FOOD AND BEVERAGE EQUIPMENT

No Cost Opportunities

- Tauranga-based Coldstorage International (now Fonterra, Mount Maunganui Cold Store) found that by measuring the product temperature in freezers (which was remarkable stable) verses the air temperature, they where able to switch off the refrigeration plant when energy costs were the highest. Emprove, New Zealand, reported that reducing the refrigeration daytime run hours saved Fonterra over 1.2 million-kilowatt hours per year.
- Don't overload fryer baskets beyond the recommended capacity. Overloading increases cook time.
- Use cooking and cleaning equipment to capacity. Fully loaded equipment utilizes energy more efficiently. Filling dishwashers to capacity can save approximately $200 annually.

- Turn off back-up fryers and ovens during low production periods. Turning off an idle broiler just one hour a day can save over $400 annually.

- Turn exhaust hoods whenever possible. Not only does the exhaust hood use energy, the air it exhausts must be replaced with outside air that may require heating or cooling. An exhaust hood can use $600 or more of energy per year.

- Keep refrigerators full. A full refrigerator retains cold better than an empty one. If the refrigerator is nearly empty, store water-filled containers inside. The mass of cold items will enable the refrigerator to recover more quickly after the door has been opened. On the other hand, don't overfill it, since that will interfere with the circulation of cold air inside.

- Regularly defrost manual-defrost models. Frost buildup increases the amount of energy needed to keep the motor running.

Low Cost Opportunities

- Add strip curtains to refrigerated spaces and display cases, or spring loaded doors that automatically shut. Even if the display case cooling is used to cool the store, extra energy is required to defrost the display cases when humidity enters the refrigerated display cases.

- Make sure oven and refrigerator doors fit tightly and gaskets are in good condition. Replacing damaged door gaskets can save $100 annually.

- Replace old, high-volume kitchen sprayers with high-velocity, low-flow models, and save up to $1,000 a year in hot water costs.

- Consider reducing the temperature of your hot water heater. 110 degrees is required for hand washing. Typically, 130 degrees is required for washing dishes.

- Purchase insulated cooking equipment whenever possible (e.g., fryers, ovens, coffee machines). Insulation retains more heat in the equipment.

- Install an insulated dishwasher. Replace a 10 to 20-year-old conventional dishwasher with an insulated model and save up to $500 annually. Insulated tanks reduce stand-by or idle energy consumption. A booster heater spends several hours per day consuming energy in a "ready-to-use" mode. Insulated tanks can better maintain an outgoing sanitizing water temperature at all times, thus, reducing stand-by energy consumption.

LAUNDRY EQUIPMENT

No Cost Opportunities

- Clean the lint from the clothes dryer after every load. The efficiency of the dryer goes down when lint collects over the dryer filter. Run full loads and use the moisture-sensing setting.

- Keep the clothes dryer's outside exhaust clean. A clogged exhaust lengthens drying time and increases energy use.

Low Cost Opportunities

- Install high-efficiency commercial washers, including but not limited to front-loading machines, which can cut energy costs up to 50% and use about 30% less water (18 to 25 gallons of water per load, compared to 40 gallons used by a standard machine). Energy-efficient and front-loading commercial clothes washers also last five to 10 years longer than standard, top-loading machines.

TRANSPORTATION

Low Cost Opportunities

For each gallon of gas you burn, 20 pounds of CO_2 is released into the atmosphere. Some simple steps can add up.

- Keep the tires of your car or truck properly inflated. Under-inflation shortens the life of a tire and decreases gas mileage. For every pound per square inch (psi) below the proper level, there is an average increase in fuel consumption of approximately 0.4 percent.

- Don't speed. For every mile-per-hour over 55 mph, the average car or truck loses almost two percent in gas mileage.

- When driving on the highway, use your cruise control to maintain a steady speed. Don't brake suddenly or accelerate quickly.

- Keep your car or truck well tuned. A well-tuned car uses up to 9 percent less fuel than a poorly tuned car and releases less pollution.

- Remove unnecessary heavy items from your car. Every extra 100 pounds can cost you about half-a-mile-per-gallon.

- Reduce drag when possible. Drag increases fuel consumption. If you drive with the windows open, more drag is created. Roof-mounted racks can increase drag by more than 40 percent if you stack luggage, bicycles, or skis on the roof and back of the car. At high speeds, it is often cheaper to run the car air conditioner than roll down the windows.

- Ride a bike or walk to work, the local neighborhood store, or nearby friends. Total vehicle emissions are then reduced to zero.

- If your car is equipped with overdrive, be sure to use the overdrive gear when your speed dictates. Your owner's manual will give you further information.

- Don't start your car until you're ready to go. Idling engines waste gas. Limit car warm-ups in winter.

- Drive smoothly. Accelerating slowly from a full stop can save you as much as two miles per gallon.

- Drive a friend or neighbor to work. If every commuter car carried just one more passenger, we'd save 600,000 gallons of gasoline and keep 12 million pounds of greenhouse gases out of the atmosphere every day.

WATER

No Cost

- Don't run water continuously when not needed.

- Water landscaping only when necessary, and then in the evenings or early morning.

- Verify sewer charges and ensure that water not discharged into the sewer system is not billed.

- Chose native, low water, plants for landscaping.

Low Cost

- Install low-flow faucet aerators to faucets.

- Install low-flow showerheads and toilets.

- Insulate hot water pipes.

- Capture and use roof runoff to irrigate the landscaping.

- Capture condensate water from chillers and use to irrigate the landscaping.

Chapter 11

Energy Awareness Facts, Quotes and Tidbits

Continual learning, measurement, and benchmarking are powerful tools to maintain the motivation to reduce energy costs. Therefore, it should come as no surprise that successful energy awareness champions continually research and study energy conservation. My overall purpose in compiling additional energy facts and quotes is to provide: (1) an easy source of material for speaking or writing about energy awareness topics; (2) a storehouse of energy facts to simulate ideas, thought, and action; (3) a quick source of additional inspiration and personal motivation that can be referenced periodically as your motivation wanes; (4) to expand on the concepts presented in Section I in a quick and easy to read format.

The reference information in this chapter is not meant to be read in one sitting, but occasionally to reinforce the message. Try to keep the quotes and energy information easily accessible and refer to them prior to speaking or writing about energy awareness topics.

ENVIRONMENTAL FACTS AND QUOTES

Knowledge of the environmental consequences of energy consumption can stimulate energy saving behavior. This inherent "environmental goodwill" drives positive energy saving actions and can be a powerful motivator. Using environmental messaging in your energy awareness initiative will reduce energy costs.

GLOBAL WARMING, IS IT FACT OR FICTION?

Atmospheric carbon dioxide has already increased to 380 parts per million (ppm), up from 280 ppm before the industrial revolution. How much higher can we go before the altered atmosphere dramatically changes weather patterns? Unfortunately, a replicable experiment to find this answer in advance is impossible; we can't go find another similar planet and raise the carbon levels until that planet experiences weather anomalies. We must rely on past carbon dioxide and temperature colorations and computer models to validate any global warming hypothesis. So, it is not surprising that there are differences in opinion among scientists; some think carbon levels as low as 400 ppm will cause drastic changes. The consensus opinion among scientists that believe in global warming is probably somewhere between 500 and 550 ppm. Since the carbon dioxide concentration in our atmosphere is increasing at about 2 ppm per year, it appears we will find out. Here is what others think:

- Long-term observations confirm that our climate is now changing at a rapid rate. Over the 20th century, the average annual US temperature has risen by almost 1°F (0.6°C) and precipitation has increased nationally by 5 to 10%, mostly due to increases in heavy downpours. These trends are most apparent over the past few decades. The science indicates that the warming in the 21st century will be significantly larger than in the 20th century. Scenarios examined in this Assessment, which assume no major interventions to reduce continued growth of world greenhouse gas emissions, indicate that temperatures in the US will rise by about 5-9°F (3-5°C) on average in the next 100 years, which is more than the projected global increase. This rise is very likely to be associated with more extreme precipitation and faster evaporation of water, leading to greater frequency of both very wet and very dry conditions. —National Assessment Synthesis Team, an advisory committee chartered under the Federal Advisory Committee Act, "CLIMATE CHANGE IMPACTS ON

THE UNITED STATES: The Potential Consequences of Climate Variability and Change."

- The scientific community has reached a strong consensus regarding the science of global climate change. The world is undoubtedly warming. This warming is largely the result of emissions of carbon dioxide and other greenhouse gases from human activities including industrial processes, fossil fuel combustion, and changes in land use, such as deforestation. Continuation of historical trends of greenhouse gas emissions will result in additional warming over the 21st century, with current projections of a global increase of 2.5°F to 10.4°F by 2100, with warming in the U.S. expected to be even higher. This warming will have real consequences for the United States and the world, for with that warming will also come additional sea-level rise that will gradually inundate coastal areas, changes in precipitation patterns, increased risk of droughts and floods, threats to biodiversity, and a number of potential challenges for public health. —"Global Warming Basics," Pew Center website

- Global warming is one of the most serious challenges facing us today. To protect the health and economic well-being of current and future generations, we must reduce our emissions of heat-trapping gases by using the technology, know-how, and practical solutions already at our disposal. —Union of Concerned Scientists

- Study of the global climate is one of humankind's most complex scientific endeavors. While much is known, there is a great deal that remains uncertain. Experts know that atmospheric levels of heat-trapping gasses have increased, and the Earth is warming faster than we have ever seen. Most believe that pollution is responsible for both these trends. There are key areas about which there is no doubt. First, certain gases trap heat within the Earth's atmosphere. Second, pollution from cars, power plants, and other sources are increasing the amount of these gases in the atmosphere, leading to their highest concentration in more than 400,000 years. Third, the additional heat trapped by this global warming pollution will, according to the laws of physics, manifest itself as additional energy driving the Earth's climate system; there is simply nowhere else for it to go. We also know that the planet is getting warmer. Global average temperatures on the Earth's surface have increased by 1.1 degrees Fahrenheit (0.6 degrees Celsius) during the last century—warming faster than any time in the last 1000 years. As a result, the 1990s was the warmest decade in the last 1000 years. —Natural Resources Defense Council, "FEELING THE HEAT IN FLORIDA"

- Humanity's influence on the global climate will grow in the coming century. Increasingly, there will be significant climate-related changes that will affect each one of us. We must begin now to consider our responses, as the actions taken today will affect the quality of life for us and future generations. —National Assessment Synthesis Team, an advisory committee chartered under the Federal Advisory Committee Act, "CLIMATE CHANGE IMPACTS ON THE UNITED STATES: The Potential Consequences of Climate Variability and Change"

- We accept that provisional or not, the science on global warming is for the present overwhelming. We believe that there should be mandatory carbon constraints. We support small, increasing ratcheting limits. We support higher energy efficiency standards in both buildings and appliances. —John Rowe, Chairman and CEO, Exelon Corporation (2004)

- We can't ignore mounting scientific evidence on important issues such as climate change. The science may be provisional. All science is provisional. But if you see a risk you have to take precautionary action just as you would in any other aspect of business. —Sir John Browne, CEO of BP

- So, is it really worth using public resources now to avert an uncertain, distant risk, especially when the cash could have been spent instead on goods and services that would have a measurable near-term benefit? If the risk is big enough, yes. Governments do it all the time. They spend a small slice of tax revenue on keeping standing armies not because they think the countries are in imminent danger of invasion, but because if it happens, the consequences would be catastrophic. Individuals do so too. They spend a little of their incomes on household insurance not because they think their homes are likely to be torched next week but, because if it happened, the results would be disastrous. Similarly, a growing body of scientific evidence suggests that the risk of a climatic catastrophe is high enough for the world to spend a small proportion of its income trying to prevent one from happening. —*The Economist,* "The heat is on"

- Most attribution studies find that, over the last 50 years, the estimated rate and magnitude of global warming due to increasing concentrations of greenhouse gases alone are comparable with or larger than the observed warming. —J.F.B. Mitchell, D.J. Karoly, G.C. Hegerl, F.W. Zwiers, M.R. Allen, J. Marengo, Detection of Climate Change and Attribution of Causes. In: *Climate Change 2001: The Scientific Basis. Contribution of Working Group I to the Third Assessment Report of the Intergovernmental Panel on Climate Change* [Houghton, J.T., Y. Ding, D.J. Griggs, M. Noguer, P.J. van der Linden, X. Dai, K. Maskell, and C.A. Johnson (eds.)]. Cambridge University Press, Cambridge, United Kingdom and New York, NY, USA, 881pp.

- Today's CO_2 levels are the highest that they have been in the past one million years; indeed, probably the highest they have been since 50mn years ago—the hottest period in the earth's history when there was no ice at the Antarctic. —Sir David King, UK government's Chief Scientific Adviser

- In the past few centuries, atmospheric carbon dioxide has increased by more than 30 percent, and virtually all climatologists agree that the cause is human activity, predominantly the burning of fossil fuels and, to a considerable extent, land uses such as deforestation. —Stephen H. Schneider, Stanford University

- Over the last century we have seen a rise in the atmospheric concentration of carbon dioxide from 280 ppm to some 370 ppm. Coincident with this rise has been an increase in global average temperature, up by nearly 1°C. Projections show that if this trend continues, global temperatures could rise by a further one to four degrees by the end of the 21st century. —World Business Council for Sustainable Development, "Facts and trends to 2050, Energy and climate change"

- The climate crisis may at times appear to be happening slowly, but in fact it is a true planetary emergency. The voluminous evidence suggests strongly that, unless we act boldly and quickly to deal with the causes of global warming, our world will likely experience a string of catastrophes, including a deadlier Hurricane Katrina in both the Atlantic and Pacific. —Al Gore, "The Moment of Truth," *Vanity Fair*

- The bottom line of the global warming, greenhouse effect issue is that we insult the environment at a faster rate than we understand the consequences. Simple prudence suggests that modifying the global climate at 10 to 50 times the average natural rates of change is not a planetary experiment that we should glibly allow, particularly since there are so many measures available that could substantially slow down our impact on Earth and at the same time buy many other benefits. —Dr. Stephen Schneider

- Increased warming is projected for the 21st century—Assuming continued growth in world greenhouse gas emissions, the primary climate models drawn upon for the analyses carried out in the U.S. National Assessment projected that temperatures in the contiguous United States will rise 3–5°C (5–9°F) on average during the 21st century. A wider range of outcomes, including a smaller warming, is also possible. —U.S. Department of State, "U.S. Climate Action Report 2002"

- Enough is known about the science and environmental impacts of climate change for us to take actions to address its consequences. —American Electric Power

- One day we will live in a carbon-constrained world. —Jim Rodgers, CEO of Cinergy Corp

- In fact, the Earth is always experiencing either warming or cooling. But suppose the scientists and their journalistic conduits, who today say they were so spectacularly wrong so recently, are now correct. Suppose the Earth is warming and suppose the warming is caused by human activity. Are we sure there will be proportionate benefits from whatever climate change can be purchased at the cost of slowing economic growth and spending trillions? Are we sure the consequences of climate change—remember, a thick sheet of ice once covered the Midwest—must be bad? Or has the science-journalism complex decided that debate about these questions, too, is "over"? —George Will, "Let Cooler Heads Prevail," *WashingtonPost.com*

- The facts are unpleasant realities:
  — We are a nation at war.
  — Our earth is warming.
  — Carbon emissions and greenhouse gases are impacting air quality and the environment.
  — America is addicted to oil.

—Alexander Karsner,
U.S. Assistant Secretary for
Energy Efficiency and Renewable Energy

- My Opinion:
  - — Climate is changing (global warming)
  - — Human (greenhouse gas) role is probable
  - — Global warming increases hydrologic extremes (droughts / fires and heavy rain / floods)
  - — With large climate change, detrimental effects probably exceed beneficial ones
  - — Common sense steps to limit climate change are warrant-ed

—Dr. James Hansen, NASA

- Several additional lines of evidence confirm that the recent and continuing increase of atmospheric CO_2 content is caused by anthropogenic CO_2 emissions—most importantly fossil fuel burning. First, atmospheric O_2 is declining at a rate comparable with fossil fuel emissions of CO_2 (combustion consumes O_2). Second, the characteristic isotopic signatures of fossil fuel leave their mark in the atmosphere. Third, the increase in observed CO_2 concentration has been faster in the northern hemisphere, where most fossil fuel burning occurs. —I.C. Prentice, G.D. Farquhar, M.J.R. Fasham, M.L. Goulden, M. Heimann, V.J. Jaramillo, H.S. Kheshgi, C. Le Quéré, R.J. Scholes, D.W.R. Wallace, 2001: The Carbon Cycle and Atmospheric Carbon Dioxide. In: *Climate Change 2001: The Scientific Basis. Contribution of Working Group I to the Third Assessment Report of the Intergovernmental Panel on Climate Change* [Houghton, J.T., Y. Ding, D.J. Griggs, M. Noguer, P.J. van der Linden, X. Dai, K. Maskell, and C.A. Johnson (eds.)]. Cambridge University Press, Cambridge, United Kingdom and New York, NY, USA, 881pp.

- Climatic data for the United States reveal changes and variations that may be significant in redistributing vector-borne and water-borne diseases, as well as direct climate-induced human morbidity and mortality. Since the turn of the century average daily temperatures in the conterminous United States have increased by approximately 0.4 degrees C, with most of this increase occurring during the past 30 years (Karl et al. 1995b). Recent studies have shown that the hydrologic cycle in the US is changing as indicated by increases in cloud cover (Karl and Steurer 1990) and precipitation (Groisman and Easterling 1994) and decreases in pan evaporation (Peterson 1996). Extremes in US precipitation have been changing with increases in heavy precipitation events and decreases in lighter precipitation events (Karl et al. 1995a; Karl et al. 1996). Using data back to 1910, Karl et al. found that the most recent 15 years had the highest values of Greenhouse Climate Response Index (GCRI) as well as Climate Extremes Index (CEI). It is becoming increasingly apparent that measurable changes in climate trends are occurring in the US. —Climate Analysis web site at Johns Hopkins University

- The world is changing faster than anyone expected. Not only is the earth warming, bringing more intense storms and causing Arctic ice to vanish, but the political and policy landscape is being transformed even more dramatically. Already, certain industries are facing mandatory limits on emissions of carbon dioxide and other greenhouse gases in some of the 129 countries that have signed the Kyoto Protocol. —"The Race Against Climate Change"

- In the three days following September 11, the entire U.S. jet fleet was grounded, over which time climatologists noted an unprecedented increase in the daytime temperatures relative to nighttime temperatures. This resulted, they presume, from the additional sunlight reaching the ground in the absence of [jet] contrails. —Tim Flannery, *The Weather Makers*

Effects of Global Warming

Weather and climate have a profound influence on our lives, but can a few extra degrees really be that bad? Are we like frogs in a simmering pot, unaware that temperatures are rising?

- An increase of a few degrees won't simply make for pleasantly warmer temperatures around the globe. Even a modest rise of 2°- 3°F (1.1°-1.7°C) could have dramatic effects. In the last 10,000 years, the Earth's average temperature hasn't varied by more than 1.8°F (1.0°C). Temperatures only 5°-9°F cooler than those today prevailed at the end of the last Ice Age, in which the Northeast United States was covered by more than 3,000 feet of ice. Scientists predict that continued global warming on the order of 2.5°-10.4°F over the next 100 years is likely to result in:

— a rise in sea level between 3.5 and 34.6 in. (9-88 cm), leading to more coastal erosion, flooding during storms, and permanent inundation

— severe stress on many forests, wetlands, alpine regions, and other natural ecosystems

— greater threats to human health as mosquitoes and other disease-carrying insects and rodents spread diseases over larger geographical regions

— disruption of agriculture in some parts of the world due to increased temperature, water stress, and sea-level rise in low-lying areas such as Bangladesh or the Mississippi River delta.

—Union of Concerned Scientists

- Twenty percent of the Arctic polar ice cap has melted since 1979 and at least half of the ice cap is projected to melt by the end of this century, along with a significant portion of the Greenland Ice Sheet, as the Arctic region warms an additional 7 to 13 degrees Fahrenheit by 2100. Melting ice sheets and glaciers threaten low-lying coastal areas, including our most heavily-populated coastal cities, with rising sea levels, deadly storm surges and salt water intrusion into drinking water supplies. —Natural Resources Defense Council

- Ecosystems are especially vulnerable—Many ecosystems are highly sensitive to the projected rate and magnitude of climate change, although more efficient water use will help some ecosystems. A few ecosystems, such as alpine meadows in the Rocky Mountains and some barrier islands, are likely to disappear entirely in some areas. Other ecosystems, such as southeastern forests, are likely to experience major species shifts or break up into a mosaic of grasslands, woodlands, and forests. Some of the goods and services lost through the disappearance or fragmentation of natural ecosystems are likely to be costly or impossible to replace. —U.S. Department of State, "U.S. Climate Action Report 2002"

- Increased damage occurs in coastal and permafrost areas—Climate change and the resulting rise in sea level are likely to exacerbate threats to buildings, roads, power lines, and other infrastructure in climate-sensitive areas. For example, infrastructure damage is expected to result from permafrost melting in Alaska and from sea level rise and storm surges in low-lying coastal areas. U.S. Department of State, U.S. "Climate Action Report 2002"

- Adaptation determines health outcomes—A range of negative health impacts is possible from climate change. However, as in the past, adaptation is likely to help protect much of the U.S. population. Maintaining our nation's public health and community infrastructure, from water treatment systems to emergency shelters, will be important for minimizing the impacts of water-borne diseases, heat stress, air pollution, extreme weather events, and diseases transmitted by insects, ticks, and rodents. U.S. Department of State, "U.S. Climate Action Report 2002"

- Increased warming is projected for the 21st century—Assuming continued growth in world greenhouse gas emissions, the primary climate models drawn upon for the analyses carried out in the U.S. National Assessment projected that temperatures in the contiguous United States will rise 3–5°C (5–9°F) on average during the 21st century. A wider range of outcomes, including a smaller warming, is also possible. —U.S. Department of State,

"U.S. Climate Action Report 2002," Washington, D.C., May 2002.

- Natural ecosystems appear to be the most vulnerable to the harmful effects of climate change, as there is often little that can be done to help them adapt to the projected speed and amount of change. Some ecosystems that are already constrained by climate, such as alpine meadows in the Rocky Mountains, are likely to face extreme stress, and disappear entirely in some places. It is likely that other more widespread ecosystems will also be vulnerable to climate change. One of the climate scenarios used in this Assessment suggests the potential for the forests of the Southeast to break up into a mosaic of forests, savannas, and grasslands. Climate scenarios suggest likely changes in the species composition of the Northeast forests, including the loss of sugar maples. Major alterations to natural ecosystems due to climate change could possibly have negative consequences for our economy, which depends in part on the sustained bounty of our nation's lands, waters, and native plant and animal communities. National Assessment Synthesis Team, an advisory committee chartered under the Federal Advisory Committee Act, "CLIMATE CHANGE IMPACTS ON THE UNITED STATES: The Potential Consequences of Climate Variability and Change"

- We face both increased flooding and increased drought. Extended heat waves, more powerful storms, and other extreme weather events will become more common. Rising sea level will inundate portions of Florida and Louisiana, while increased storm surges will threaten communities all along our nation's coastline. New York City could face critical water shortages as rising sea level raises the salinity of upstate aquifers and reservoirs. And a good chunk of lower Manhattan that's built on landfill could again be submerged. We can adjust to some of these things, if we're willing to pay the price. But many of the projected impacts are irreversible—when we lose a fragile ecosystem like the Everglades or Long Island Sound, it can never be replaced. —"Climate Change: Myths and Realities," *Pew Center Website*

- Some of our islands are only a few meters wide in places. Imagine standing on one of these islands with waves pounding on one side and the lagoon on the other. It's frightening. —Nakibae Teuatabo, chief climate negotiator for the Pacific islands state of Kiribati, describing the impact of rising sea levels

- By one estimate, the world's glaciers lose at least 90 cubic kilometers of ice annually—as much water as all U.S. homes, factories, and farms use every four months. Scientists suspect that the enhanced melting is related to the unprecedented release of greenhouse gases by humans during the past century. —World Watch Institute

- Desertification puts some 135 million people worldwide at risk of being driven from their lands. As climate change translates into more intense storms, flooding, heat waves, and droughts, more and more communities will likely be affected. —World Watch Institute

- Some ecosystems that are already constrained by climate, such as alpine meadows in the Rocky Mountains, are likely to face extreme stress and disappear entirely in some places. —U.S. Department of State, "U.S. Climate Action Report 2002"

- The demise or continued deterioration of reefs could have profound implications for the United States. Coral reefs play a major role in the environment and economies of Florida and Hawaii as well as in most U.S. territories in the Caribbean and Pacific. They support fisheries, recreation, and tourism and protect coastal areas. In addition, coral reefs are one of the largest global storehouses of marine biodiversity, sheltering one-quarter of all marine life and containing extensive untapped genetic resources. The last few years have seen unprecedented declines in the health of coral reefs. The 1998 El Niño was associated with record sea-surface temperatures and associated coral bleaching (which occurs when coral expel the algae that live within them and that are necessary to their survival). In some regions, as much as 70 percent of the coral may have died in a single season. There has also been an upsurge in the variety, incidence, and virulence of coral diseases in recent years, with major die-offs in Florida and much of the Caribbean region (NCAG 2000). —U.S. Department of State, "U.S. Climate Action Report 2002"

- The arid American West appears to be particularly susceptible to the effects of global warming. Over the past 50 years, the western climate has warmed on average by 1.4 degrees Fahrenheit, and climate models predict a further increase of 3.6 to 12.6 degrees Fahrenheit in the West by the end of the century. A growing body of scientific evidence is linking global warming trends with changes in precipitation, declining snowpack, and smaller and earlier spring runoff—conditions that determine the quantity and timing of water supplies in the West, as well as wildfire risk. Many parts of the West are already experiencing devastating multi-year droughts. If current global warming trends continue, they present serious consequences to many bedrock elements of western life, from agriculture and ranching to skiing, tourism, biodiversity and public health. —Natural Resources Defense Council

- The midrange estimate is that 24 percent of plants and animals will be committed to extinction by 2050 —ecologist Chris Thomas of Britain's University of Leeds, *Washington Post*

- The local sea level rise expected in coastal Florida could range from 8 inches to 2.5 feet by 2100. But more important is the fact that horizontal advance of the sea can be much greater than the vertical rise. The horizontal advance can be 150 to 200 times the rise, and even more than this in areas with a gently sloping shoreline as is the case for much of Florida. As a result, sea level rise over the next century could directly cause flooding of homes, hotels, and property within 200 to 250 feet of the current shoreline. This translates into a 20 to 25 percent increase in the 100-year floodplain, which in turn may increase flood damages by 40 to 60 percent. —Natural Resources Defense Council, "FEELING THE HEAT IN FLORIDA"

- It's always devastating when you witness a species' extinction, for what you are seeing is the dismantling of ecosystems and irreparable genetic loss. The golden toad's extinction, however, was not in vain, for when the explanation of its demise was published in *Nature*, the scientists could make their point without equivocation. The golden toad was the first documented victim of global warming. We had killed it with our profligate use of coal-fired electricity and our oversized cars just as surely as if we had flattened its forest with bulldozers. —Tim Flannery, *The Weather Makers*

- The oceans have absorbed half of all the gas so far emitted by humanity, and will go on doing so. This process forms dilute carbonic acid, which hinders the ability of corals, crustaceans, mollusks and certain plankton to form their hard structures or shells. As the acidity continues to rise, it is feared, coral reefs, shellfish and plankton will die off, with huge knock-on effects on the life of the oceans. —"Heating-up," ourplanet.com

- Climate change has a range of complex interlinkages with health. These include direct impacts, such as temperature-related illness and death; the health impacts of extreme weather events; the effect of air pollution in the form of spores and moulds. Other impacts follow more intricate pathways such as those that give rise to water- and food-borne diseases; vector-borne and rodent-borne diseases; or food and water shortages. Climate change impacts will not be evenly distributed around the world. Some regions are expected to fare worse than others. Small Island States, for example, are amongst the most vulnerable. Many other developing countries are also not prepared for potential environmental impacts, and even less for health-related impacts. —"Climate Change and Health," World Health Organization website

- In terms of extreme weather events, it's worth recording that the United States already has the most varied weather of any country on Earth, with more intense and damaging tornadoes, flash floods, intense thunderstorms, hurricanes, and blizzards than anywhere else. With the intensity of such events projected to increase as our plant warms, in purely human terms the United States would seem to have more to lose from climate change than any other large nation. Indeed, its ever spiraling insurance bill resulting from severe weather events and its growing water shortages in the west mean that the United States is already paying dearly for its CO_2 emissions. —Tim Flannery, *The Weather Makers*

Global Warming Tipping Points

Thus far, climate change as been slow. Some scientists are worried that events will greatly expedite or irrevocability increase global warming. For example, should much of the polar ice and glaciers melt, and they are melting, global warming may increase faster. This is because the existing ice reflects much of the sun's heat, the land and water left behind absorbs and stores more of the sun's heat. Other factors, such as the acidly of the oceans, salinity of the oceans, changes in ocean currents, changes in plankton, or cloud formations could all reach levels that cause a cascading effect.

- The more the climate is forced to change, the more likely it is to hit some unforeseen threshold that can trigger quite fast, surprising and perhaps unpleasant changes. —Richard Alley, Professor of Geosciences at Pennsylvania State University

- Everything you thought you knew about the rainforest may soon prove completely wrong. In a remote portion of the Costa Rican jungle, a team of ecologists is measuring every known quantity about tropical forests—every piece of gunk that falls to the ground, every wisp of carbon that rises to the sky, how much air goes into the soil, what the bugs are eating. And they're finding that as the temperature rises from global warming, the rainforest—long thought to be a great repository of greenhouse gas—grows more slowly. This discovery, and the data they've collected since making it, indicate that if the rainforest's temperature crosses a certain as-yet unknown threshold, there's a very good chance the trees will begin to decay, emitting carbon instead of storing it. And we all know what that means. —Hillary Rosner, *SEED*

- If the West Antarctic ice sheet ever does detach itself from the sea floor, it would add 6 to 20 inches of sea level rise by 2100. Even worse, the glaciers feeding into it would accelerate, adding much more to sea levels. In all, the .9 million cubic miles of sea and glacial ice contained and held back by the West Antarctic ice sheet constitute enough water to raise global sea levels by 20 to 23 feet. —Tim Flannery, *The Weather Makers*

- ... the Siberian and Alaskan tundras, which for centuries absorbed carbon dioxide and methane, are now thawing and releasing those gases back into the atmosphere. A rapid release of greenhouse gases from these regions could trigger a spike in warming. —Ross Gelbspan, "Why we need to worry about global warming, now," *San Jose Mercury News*

- Models project that a local annual average warming of larger than 3°C, sustained for millennia, would lead to virtually a complete melting of the Greenland ice sheet with a resulting sea level rise of about 7 m. —IPPC working report, J.A. Church, J.M. Gregory, P. Huybrechts, M. Kuhn, K. Lambeck, M.T. Nhuan, D. Qin, P.L. Woodworth, 2001: Changes in Sea Level.

In: Climate Change 2001: The Scientific Basis. Contribution of Working Group I to the Third Assessment Report of the Intergovernmental Panel on Climate Change [Houghton, J.T., Y. Ding, D.J. Griggs, M. Noguer, P.J. van der Linden, X. Dai, K. Maskell, and C.A. Johnson (eds.)].

Paradoxes of Global Warming

I have heard people say, "We had a massive snow storm last winter, so there can't be global warming." That may not be a relevant correlation. Apparently, some aspects of global warming theory can appear contradictory.

- Global warming would also bring more heat waves like the summer of 2003 that killed 31,000 across Europe. It might even shut down the Gulf Stream, the flow of warm water from the Gulf of Mexico that gives Europe its mild climate. If the Gulf Stream were to halt—and it has already slowed 30 percent since 1992—Europe's temperatures would plunge, agriculture would collapse, London would no longer feel like New York but like Anchorage. —Mark Hertsgaard, "While Washington Slept', *Vanity Fair*

- Through time, climate change will possibly affect the same resource in opposite ways. For example, forest productivity is likely to increase in the short term, while over the longer term, changes in processes such as fire, insects, drought, and disease will possibly decrease forest productivity. —National Assessment Synthesis Team, an advisory committee chartered under the Federal Advisory Committee Act, *CLIMATE CHANGE IMPACTS ON THE UNITED STATES: The Potential Consequences of Climate Variability and Change*

- Threats such as cholera and malaria, however, may pale in comparison to the risk of food shortages caused by global warming. Although increasing concentrations of carbon dioxide might actually boost global food production in the near future, prolonged warming and precipitation shifts will probably lower agricultural yields in many developing countries, according to the IPCC. Such countries have fewer resources to adapt to changing conditions, and their climates are already marginal for certain crops. —Richard Monastersky, "How would global warming affect humans?" *ScienceNewsOnline*

- Rising global temperature means more than just extra time to wear shorts and sandals. An increase of just a few degrees in average temperature can cause dramatic changes in conditions that are important to the quality of life —and even the Earth's ability to support life. We may not always see or feel it directly, but climate change affects us all. For one person it might mean paying more for food because flooding or drought has damaged crops. For another it might mean a higher risk of contracting a disease like malaria, which spreads more easily in warm, wet climates. Someone else might face losing her home or even life in a catastrophic weather disaster made worse by global warming. —Worldwatch Institute

- An increase of two or three degrees wouldn't be so bad for a northern country like Russia. We could spend less on fur coats, and the grain harvest would go up. —Vladimir Putin, Russian President.

- New analyses show that in regions where total precipitation has increased, it is very likely that there have been even more pronounced increases in heavy and extreme precipitation events. The converse is also true. In some regions, however, heavy and extreme events (i.e., defined to be within the upper or lower ten percentiles) have increased despite the fact that total precipitation has decreased or remained constant. This is attributed to a decrease in the frequency of precipitation events. Overall, it is likely that for many mid and high latitude areas, primarily in the Northern Hemisphere, statistically significant increases have occurred in the proportion of total annual precipitation derived from heavy and extreme precipitation events; it is likely that there has been a 2 to 4% increase in the frequency of heavy precipitation events over the latter half of the 20th century. IPPC working report, "CLIMATE CHANGE 2001: THE SCIENTIFIC BASIS," CAMBRIDGE UNIVERSITY PRESS

- ...A plant breathes through small holes, called stomata, found in its leaves. Plants take in carbon dioxide, and when the atmosphere is relatively rich in this gas, less effort is needed. The stomata stay closed for longer, and less water is lost to the atmosphere. This means that the plant doesn't need to draw as much moisture from the soil. The unused water flows into rivers—"Full to bursting," *The Economist,* February 18th-24, 2006 discussing a theory by Dr. Gedney on why rivers around the world have become fuller over the past century.

Connecting Energy Consumption and Global Warming

Burning fossil fuels to generate the electricity that fuels our lights, motors, and computers is a large source of greenhouse gas emissions.

- Energy production and use account for nearly 80 percent of air pollution, more than 88 percent of heat-trapping greenhouse gas emissions, and more environmental damage than any other human activity. —Alliance to Save Energy

- Energy Consumption is also a major contributor to climate change, which is the cause of increasing concern over recent years. Energy is the source of 4/5 (78%) of total greenhouse gas emissions in the EU. Of these, the transportation sector contributes around one third. —"European Commission Green Paper"

- Energy saving is without a doubt the quickest, most effective and most cost-effective manner for reducing greenhouse gas emissions, as well as improving air quality, in particular in densely populated areas. —European Commission, *"Doing more with less"*

- Replacing an incandescent bulb with a compact fluorescent will save the energy equivalent of 46 gallons of oil as well as one-half ton of carbon dioxide emissions over the lifetime of the bulb. —U.S. Navy

- Energy reduction is of particular importance because of the emissions associated with its production and use, —Albert Young, Energy Conservation Manager

- Carbon emissions from U.S. motor gasoline use in 2002—at 1,139 million tons—surpassed those of the entire Japanese economy. —World Watch Institute

- The average carbon dioxide concentration has increased more than 19 percent since measurements began in Hawaii in 1959—and has gone up 35 percent since the dawn of the industrial age. —World Watch Institute, *Vital Signs*

- Leaving your personal computer (PC) on overnight wastes energy—and lots of it. If you leave your personal computer on overnight for a year, you are wasting electricity equivalent to what the typical U.S. power plant generates from 57 gallons of oil to drive the average American car 1,380 miles. And that's just one PC, imagine what turning off all of the PCs around the base could save! —U.S. Navy

- Climate change is driven in large part by carbon dioxide and other greenhouse gases derived from burning fossil fuels. In California, electricity production and industry are the source of nearly half of fossil fuel-related carbon dioxide. Other harmful emissions released by burning fossil fuels to produce electricity include sulfur dioxide and nitrous oxide, two of the primary sources of smog. Energy efficiency helps reduce the need to generate electricity. —California Flex Your Power Program

- Between 1978 and 1991, there was a 4-5 percent loss of ozone in the stratosphere over the United States, which represents a significant loss of ozone. A thinned-out ozone layer could lead to more skin cancers and cataracts; scientists are also investigating possible harm to agriculture. Destruction of stratospheric ozone is attributed to CFCs and related chemicals. CFCs are widely used as refrigerants in such appliances as refrigerators, freezers, air conditioners, and heat pumps. —U.S. Navy

- The science is clear and compelling: We humans are changing the global climate. Concentrations of greenhouse gases in the atmosphere are at their highest level in more than 200,000 years and climbing sharply. —President Bill Clinton

- Here in the United States we must do better. With 4 percent of the world's population, we already produce more than 20 percent of its greenhouse gases. —President Bill Clinton

- Every time a light, a computer, or an air conditioner is turned on, a power plant consumes some type of fuel to generate electricity. Most of the time, a fossil fuel is burned to generate power, and the process releases emissions and pollution into the atmosphere and elsewhere in the environment. These pollutants cause acid rain and smog, and most scientists agree these emissions are changing our global climate. Human health—as well as nature's complex ecosystem—also is affected directly. —Energy Star

- By actively pursuing energy efficiency measures we are demonstrating leadership in reducing greenhouse gas emissions—Doug Dunn, Manager Building Operations

- Carbon Dixiode (CO_2) is the biggest contributor to global warming and CO_2 is mainly emitted as the result of energy consumption. So every time you turn on your heating, switch on lights, heat water, cook, or use any gas or electrical appliance in your home, you are adding to the threat of global warming. Most of us are using much more energy than we need to and producing unnecessary CO_2 emissions. The average [UK] home uses enough energy to create 7.5 tonnes of CO_2 a year. Over a quarter of the CO_2 produced in the UK comes from energy used in the home. The figure increases to around 40% if you include the use of cars. If we cut energy use in our homes and cars, we reduce the CO_2 emissions that increase the risk of global warming. —Action Energy

- Replacing just one incandescent light bulb with a compact fluorescent lamp would save 500 pounds of coal and over 1/2 ton of CO_2 emissions. —*National Geographic*

- By all means, let us use the small input from renewables sensibly, but only one immediately available source does not cause global warming and that is nuclear energy. True, burning natural gas instead of coal or oil releases only half as much carbon dioxide, but unburnt gas is 25 times as potent a greenhouse agent as is carbon dioxide. Even a small leakage would neutralize the advantage of gas. —James Lovelock, "Nuclear Power is the only green Solution," *The Independent*

- Here in the United states, most of us begin generating CO_2 as soon as we get out of bed. Seventy percent of our electricity is generated by burning fossil fuels—a little more than 50 percent from burning coal and other 17 percent from natural gas-so that to turn on the lights is, indirectly at least, to pump carbon dioxide into the atmosphere. Making a pot of coffee, either on an electric or gas range, adds more emissions, as does taking a hot shower, watching the morning news on TV, and driving to work. —Elizabeth Kolbert, *Field Notes from a Catastrophe*

Global Warming, is it too Late to Act?

With carbon dioxide levels at record highs and climbing, can we really make a difference?

- We could save about nine out of 10 species under threat, limit the extent of extreme weather so that losses of life and investments are a fraction of those predicted and reduce, almost to zero, the possibility of any great disasters occurring this century. But for that to happen, individuals, industry and governments need to act now. A delay of even a decade is too much. —Tim Flannery, *The Weather Makers*

- In order to curb global warming, we must stabilize concentrations of greenhouse gases, such as carbon dioxide (CO_2), in our atmosphere, and we must begin immediately. Scientists have warned that, even in order to stabilize CO_2 concentrations at a level twice as high as pre-industrial levels, global emissions must reverse course by 2013, only 11 years from today. Yet, U.S. CO_2 emissions continue to rise at an alarming rate. The Energy Information Administration reported in November 2001 that U.S. CO_2 emissions in 2000 were 17 percent higher in 2000 than in 1990. Electric utility emissions grew at an even higher rate, jumping 26.5 percent from 1990 to 2000. —Rebecca Stanfield, "*Darkening Skies; Trends Toward Increasing Power Plant Emissions*"

- The worse case scenarios of global warming might still be avoided, scientists say, if humanity reduces its greenhouse-gas emissions dramatically, and very soon. The I.P.C.C [Intergovernmental Panel on Climate Change] has estimated that emissions must fall to 60 percent below 1990 levels before 2050, over a period when global population is expected to increase by 37 percent and per-capita energy consumption will surely rise as billions of people in Asia, Africa, and South America strive to ascend from poverty. Yet even if such as reduction were achieved, a significant rise in sea levels may be unavoidable. —Al Gore, "While Washington Slept"

- If greenhouse gas concentrations were stabilized, sea level would nonetheless continue to rise for hundreds of years. After 500 years, sea level rise from thermal expansion may have reached only half of its eventual level, which models suggest may lie within ranges of 0.5 to 2.0 m and 1 to 4 m for CO_2 levels of twice and four times pre-industrial, respectively. Glacier retreat will continue and the loss of a substantial fraction of the total glacier mass is likely. Areas that are currently marginally glaciated are most likely to become ice-free. Ice sheets will continue to react to climate change during the next several thousand years even if the climate is stabilized. Models project that a local annual-average warming of larger than 3°C sustained for millennia would lead to virtually a complete melting of the Greenland ice sheet. For a warming over Greenland of 5.5°C, consistent with mid-range stabilization scenarios, the Greenland ice sheet contributes about 3 m in 1,000 years. For a warming of 8°C, the contribution is about 6 M, the ice sheet being largely eliminated. For smaller warmings, the decay of the ice sheet would be substantially slower. —J.A. Church, J.M. Gregory, P. Huybrechts, M. Kuhn, K. Lambeck, M.T. Nhuan, D. Qin, P.L. Woodworth, 2001: Changes in Sea Level. In: *Climate Change 2001: The Scientific Basis. Contribution of Working Group I to the Third Assessment Report of the Intergovernmental Panel on Climate Change* [Houghton, J.T.,Y. Ding, D.J. Griggs, M. Noguer, P.J. van der Linden, X. Dai, K. Maskell, and C.A. Johnson (eds.)].

- Our house is burning down and we're blind to it... The earth and humankind are in danger and we are all responsible. It is time to open our eyes. Alarms are sounding across all the continents... We cannot say that we did not know! Climate warming is still reversible. Heavy would be the responsibility of those who refused to fight it. —French President Jacques Chirac

- Measurement of health effects from climate change can only be very approximate. Nevertheless, a WHO quantitative assessment, taking into account only a subset of the possible health impacts, concluded that the effects of the climate change that has occurred since the mid-1970s may have caused over 150,000 deaths in 2000. It also concluded that these impacts are likely to increase in the future. —World Health Organization, "Climate and health, Fact sheet"

Green Power

As the price of oil rises, alternative fuels become economically viable. A small, but growing number of companies and consumers pay a premium to use green power to reduce emissions and dependence on foreign sources. The European Union has a goal of having renewables generate 22 percent of Europe's electricity by 2010.

- Today, the world uses about 13 terawatts of power, approximately 80 percent of it from carbon-dioxide-emitting fossil fuels. If we want to keep Earth's average temperature low enough to prevent eventual sea-level rises—and also accommodate 3 percent annual growth—we will need between 10 and 30 terawatts of new carbon free power by 2050. —Marty Hoffert, "It's not to early," *Technology Review*

- Energy efficiency is a "homegrown" resource! —Alliance to Save Energy

- We Texans are so smart that we leave our huge cleaner resources of sun, wind and natural gas grossly underutilized while we buy $1 billion a year of Montana coal and bring it down by train to burn in Texas coal plants like "Big Brown" so as to turn our blue skies to brown. —Sam Wyly

- The amount of solar radiation that reaches the earth's surface in approximately 3 days equals roughly the total energy content of all known supplies of fossil fuels. —U.S. Navy

- U.S. wind farms cranked out 17 Billion kilowatt-hours of electricity in 2005 and are expected to harvest 100 billion kilowatt-hours by 2020. —*Business* 2.0, March 2006

- Farmers can earn money by leasing the wind rights to their land, which can bring in an estimated $4,000 a year per turbine—or by investing in the turbines and selling power to utilities for about $7,500 a year. —*Business* 2.0, March 2006

- Ultimately, the world will need to get most of its energy from renewable sources. Besides biomass, a number of other promising candidates exist. But as the demand for energy increases, we also need a long term strategy to create truly disruptive alternatives. —*Technology Review*, July/August 2006

Mercury

Mercury emissions are less controversial than carbon dioxide. Few deny that ingesting large amounts of mercury is detrimental to our health.

- U.S. electric utilities release approximately 48 tons of mercury every year. In late 2000, the U.S. Environmental Protection Agency (EPA) announced that it would regulate electric power industry mercury emissions. In December 2003, the agency proposed rules to regulate mercury from new and existing coal-based power plants and nickel from oil-based power plants. On March 15, 2005, EPA finalized a rule to regulate mercury from coal-based power plants. —Edison Electric Institute, "STRAIGHT ANSWERS About Electric Utilities and Mercury"

- When an American hockey player suffered symptoms from mercury contamination, he never expected that he might have power plants half way across the world in China to blame. With its growing appetite for energy, China is finding its many coal-burning power plants hard at work generating the much needed electricity power—as well as huge amounts of air pollutants like sulfur dioxide and mercury. The earth's climate system, however, does not recognize national borders, and that is how increased quantities of Chinese pollutants have joined, what the authors in this article call, a global "conveyor belt of bad air." This conveyor belt circles around the world, sending airborne polluting chemicals and particulates from one country to another, posing global health threats. Some scientists have estimated that 30% or more of the mercury settling into America's ecosystems comes from abroad—China, in particular. Experts have sought for possible solutions to transboundary pollution, but international environmental treaties have seldom worked effectively and the economic incentive for Chinese power plants is weak in this case. How to make local policies responsive to global problems?—YaleGlobal

- Mercury emissions from coal-fired power plants and other industrial sources are making the fish in our lakes, rivers, and streams unsafe to eat. Coal fired power plants are by far the nation's largest unregulated source of mercury emissions, contributing 41 percent of all U.S. mercury emissions. The mercury deposits in soil and surface waters, where bacteria convert it to a highly toxic form of mercury that bioaccumulates in fish, including popular sport and commercial fish. This report analyzes new data from the U.S. Environmental Protection Agency (EPA) to determine the extent to which fish in the nation's lakes are contaminated with mercury. —Emily Figdor, "Reel Danger: Power Plant Mercury Pollution and the Fish We Eat," August 2004

- Despite the importance of fish in the diet, it is nonetheless hard to overlook the mercury problem as more countries conduct tests showing extensive mercury contamination in their populations. Experts estimate that almost half (44 percent) of young children in France and 630,000 babies born each year in the United States, for example, have mercury levels exceeding health standards and are at risk of mercury poisoning. —Linda Greer, Michael Bender, Peter Maxson, and David Lennett, "Curtailing Mercury's Global Reach," The Worldwatch Institute, *State of the World 2006*

General Climate Quotes for an Awareness Program

It is easy to forget the environmental implications of simple everyday choices. Yet reminders of such choices are a key element to reducing energy costs.

- The Supreme Reality of Our Time is... the Vulnerability of our Planet. —John F. Kennedy

- It is often necessary to make decisions on the basis of information sufficient for action, but insufficient to satisfy the intellect. —Immanuel Kant

- If we do it right, protecting the climate will yield not costs, but profits; not burdens, but benefits; not sacrifice, but a higher standard of living. —Bill Clinton

- While we have taken these actions and policy positions because they are the right things to do, I want to stress that there has also been a strong business case for what we have done. In working to reduce greenhouse gas emissions, we achieved more than $2 billion in avoided costs due to energy conservation activities—and that was before the significant energy price increases of the last few years. —Chad Holliday, Chairman and CEO, DuPont

- Climate is an angry beast and we are poking at it with sticks. —Wallace Broeker

- Nature provides a free lunch, but only if we control our appetites. —William Ruckelshaus

- The only thing we have to fear on this planet is man. —Carl Jung
- Conservation of energy also protects our environment. —Lamar S. Smith
- We stand today poised on a pinnacle of wealth and power, yet we live in a land of vanishing beauty, of increasing ugliness, of shrinking open space and of an overall environment that is diminished daily by pollution and noise and blight. This, in brief, is the quiet conservation crisis. —Stewart L. Udall
- The moment one gives close attention to anything, even a blade of grass, it becomes a mysterious, awesome, indescribably magnificent world in itself. —Henry Miller
- A Healthy Ecology is the Basis for a Healthy Economy —Claudine Schneider
- We have met the enemy and they are us. —Walt Kelly
- After one look at this planet any visitor from outer space would say 'I want to see the manager'. —William S. Burroughs

FINANCIAL FACTS AND QUOTES

The energy savings potential of simple employee efforts are difficult to quantify, but are surprising in magnitude. Therefore, it is often helpful to reference other similar organizations to gauge the potential savings. This section contains additional financial information that is not included in Chapter 6, so be sure to also occasionally review Chapter 6 when looking for financial savings examples.

Energy Savings Examples to Garner Support

- Through elementary energy efficiency steps BT, the UK telecommunications group, saved $214M between 1991 and 2004. —*Financial Times*

- Through elementary energy efficiency steps DuPont, the chemicals group, has saved up to $2bn since 1990. —*Financial Times*

- Bank of America reduced its overall electrical use by an estimated 15 million kWh in 2001 as a result of retrofits and energy conscious employee actions, such as turning off lights and signage. The company's Power Down conservation initiative alone contributed savings of 13 million kWh and $1.65 million. Employee outreach, which cost $83,000, contributed the most savings. —*California Flex Your Power*

- The team lead by Brookfield LePage Johnson Controls (BLJC) together with Canada Post and BC Hydro, developed an energy and environmental conservation awareness program (EECAP) for Canada Post employees. As a result of the program, Canada Post employees participated in the program resulting in an annual energy savings of $92,000. We are extremely pleased with the results and plan to expand the program even further. —Hitesh Tailor, Regional Manager—Technical Brookfield LePage Johnson Controls Facility Management Services Ltd as posted on the BC Hydro website.

- Employee action helped ACSA [Automobile Club of Southern California] consume 2,709000 kWh less in the summer of 2001 compared with the summer of 2000. For the entire year, ACSC reduced energy consumption by 11.8 percent compared with 200 usage—a saving of 4.8 megawatts (MW). ACSC's conservation contest costs $1,000 in award money, but the returns on the investment were tremendous. —*Flex Your Power Program*

- This [energy awareness] project was initiated as part of an on-going management development program, and the costs were justified on the basis of its success as a team building exercise. The energy cost savings achieved as a result of the awareness campaign were therefore a considerable 'bonus'. Nevertheless, the total costs of £3,500 for the [management] team-building course, and £250 to publish energy awareness material, were recovered in less than 20 weeks and the savings are expected to continue. —Farley Health Products reported by Action Energy

- 86% of cost savings [were] achieved with no-cost measures and 30% reduction in waste to landfill. —Kingsmead Carpets reported by Action Energy

- The [energy awareness] program was successful in raising the level of energy awareness within the factory: many employees asked questions about energy use and what they could do to make savings. A significant improvement in the level of energy housekeeping was noted throughout the site. Action taken included consistently switching off conveyor belts and machines during production breaks. These changes resulted in an overall reduction of 3.5% in site electricity consumption, worth £10,000/year. —Farley Health Products reported by Action Energy

- An accountant highlighted £2,000 of overtime in one month in a large energy intensive industry, but did not mention that the monthly energy bill at £50,000, amounted to 25 times the overtime paid. When this fact came to light, the production department counted energy instead of overtime and costs fell by £13,000 per month—enough to pay for the entire accounts department! —Action Energy

- Colorado School District 11 developed a resource conservation management program that provides a cash incentive award to schools based on their student population and measured energy savings. In response schools have developed educational programs to raise awareness among teachers, students, and staff and encourage them to save energy. To date, District 11 has achieved total energy cost savings of almost $4 million, including more than $750,000 in 2004 alone, and awarded $329,000 back to schools for their participation. —U.S. Environmental Protection Agency Web Site

The Saving Potential of Simple Energy Reduction Actions

- We currently waste about half of the energy we use in this country, so the potential [for energy savings] is enormous. —Kateri Callahan, President the Alliance to Save Energy

- An action as simple as turning off unnecessary lights is the most effective cost-saving measure available. Every dollar saved on building operation is more money that can be spent on patient care [in hospitals.] " Engineering and Maintenance," *KGH Spectrum*, Vol 7, No. 21, November 23, 1998, as quoted by Natural Resources Canada, *Turn Energy Dollars into Health Care Dollars, March 2003.*

- In many organizations, there is enormous potential to save energy through no cost measures by raising the awareness and motivation levels of staff who are end users of energy. Achievable savings can be in order of 10%-15% of the annual energy bill. It is vital to encourage end-users of energy to avoid wasting and to take personal initiates to save energy. But changing attitudes, behavior and habits can be difficult. —John Mulholland, "Motivating People to Save Energy"

- Shut –off opportunities: this is often the most intuitive of all diagnostic procedures. However, the use of whole building data, even with heating and cooling removed can cause some confusion, since night-time electric use in many buildings is 30%-70% of daytime use. If night-time and week-end use seems high, then the connected load must be investigated to determine whether observed consumption patterns correspond to reasonable operating practices. Our experience indicated that while many if not most opportunities for equipment shut-off by an EMCS [energy management and control system] or other system-level action have been implemented, time series data analysis can still find opportunities in 10-20% of buildings. —David E. Claridge, Mingsheng Liu, and W.D. Turner

- The rates of returns from investments to improve an organization's energy performance far exceed those of many other investment options. —Energy Star

- Companies that invest in cost-effective energy-efficient products and services can lower their electric bills and enhance corporate profitability. Investments in efficiency technologies typically have a payback period of 1 to 8 years. —California Flex Your Power Program

- Consumers and businesses can also lower the up-front cost of buying energy-efficient products by taking advantage of rebates and other incentives available through your utility company, water agency and other organizations. —California Flex Your Power Program

- Between 1973 and 1986 the U.S. economy grew by 36 percent with no increase in energy use. If Americans had not become more energy efficient, annual energy bills would have been $150 billion higher. —U.S. Navy

- If over the next 15 years everyone were to buy only those energy-efficient products marked in store with EPA's distinctive Energy Star label, we could shrink our energy bills by a total of about $100 Billion over the next 15 years and dramatically cut greenhouse gas emissions. —President Bill Clinton (1997)

- Despite impressive achievements so far, America still wastes upwards of $300 billion a year worth of energy: more than the entire military budget, far more than the federal deficit, and enough to increase personal wealth by more than $1,000 per American per year. —Amory B. Lovins

- To cut your energy bills by 30 percent, look for the Energy Star label, the symbol for energy efficiency, when shopping for room air conditioners, major appliances, lighting, windows, home electronics, and office equipment. —Energy Star

- Every business uses energy—most could use less. Experience shows that energy costs can usually be reduced by at least 10%, and often by as much as 20%, by simple actions that produce quick returns. —Action Energy

- We estimate that one workstation (computer and monitor), if left on after business hours and without automatic power management, produces nearly one ton of CO_2 per year. This is five times the amount produced if the workstation is switched off at night and engages power management during idle periods in the day. If everyone in the U.S. were to turn off their equipment at night, the nation could shut down eight large power stations and save 7 million tons of CO_2 every year. One estimate (Koomey et al., 1995) is that without power management, electricity for PCs and monitors would cost U.S. businesses about $1.75 billion per year in the year 2000. (Based on a rough national average 1.35 lbs CO_2/kWh). —Roger E. Picklum, Bruce Nordman, Barbara Kresch, "Guide to Reducing Energy Use in Office Equipment"

- While not a direct use of electricity, the paper used in office equipment (copiers, printers, and fax machines) is similar in many ways to office equipment electricity use, and so a logical part of a comprehensive program. The average office worker uses 10,000 sheets per year of office paper, which is the energy equivalent of 80 Watts of electricity used during work hours, or 160 kWh/year. At half a cent per sheet, it is $50 per person per year. Significant cost-effective opportunities exist to reduce this. —Roger E. Picklum, Bruce Nordman, Barbara Kresch, "Guide to Reducing Energy Use in Office Equipment"

- The average household spends $1,400 each year on energy bills. By choosing Energy Star-qualified products, consumers can cut this by 30 percent, saving about $400 each year. —Energy Star.

Energy Forecasts

- At 6.4 billion and climbing, the world's population is expected to exceed 9 billion by 2050. Yet our known fossil fuel reserves are in decline, and alternative energy sources are not expanding rapidly enough to meet future demand. —Chevron website

- Two hundred and forty-nine coal-fired power plants are projected to be built worldwide between 1999 and 2009, almost half of which will be in China. A further 483 will follow in the decade to 2019, and 710 more between 2020 and 2030. About a third of these will be Chinese, and in total they will produce 710 gigawatts (710,000 megawatts) of power. —Tim Flannery, *The Weather Makers*

- Emerging economies account for much of the projected growth in marketed energy consumption over the next two decades, with energy use in the group more than doubling by 2025. Strong projected economic growth drives the demand for energy use in the region. Economic activity, as measured by gross domestic product (GDP) in purchasing power parity terms, is expected to expand by 5.1 percent per year in the emerging economies, as compared with 2.5 percent per year in the mature market economies and 4.4 percent per year in the transitional economies of eastern Europe and the former Soviet Union (EE/FSU). —Energy Information Administration, *International Energy Outlook 2005*

- Natural gas is projected to be the fastest growing primary energy source worldwide, maintaining average growth of 2.3 percent annually over the 2002 to 2025 period. Total world natural gas consumption is projected to rise from 92 trillion cubic feet in 2002 to 128 trillion cubic feet in 2015 and 156 trillion cubic feet in 2025. —Energy Information Administration, *International Energy Outlook 2005*

- Coal use worldwide is projected to increase by 2.0 billion short tons between 2002 and 2015 and by another 1.0 billion short tons between 2015 and 2025. In this year's outlook for coal, all regions of the world show some increase in coal use, except for Western Europe, where natural gas and, to a lesser extent, renewable energy sources are increasingly being substituted for coal to fuel electric power generation. On a regional basis, slightly lower coal use is anticipated relative to last year's outlook in the mature market economies. In the transitional economies of the EE/FSU region, coal use was expected to decline somewhat in the *IEO2004* forecast, but in this year's forecast it is expected to increase by 0.5 percent per year between 2002 and 2025. —Energy Information Administration, *International Energy Outlook 2005*

- For all sectors, demand for electricity is projected to grow more rapidly than direct fuel use in other sectors, as electricity assumes an expanding role in meeting the energy demands of the U.S. economy. Emissions of CO_2 from the electricity sector are projected to rise by 34.9 percent over the 20-year projection period. Efficient production and use of electricity, as well as development of clean fuels, will be a continuing policy focus for the United States. U.S. Department of State, "U.S. Climate Action Report 2002"

- The business-as-usual-perspective on global energy demand is neither sustainable nor acceptable, Mr. Mandil of IEA told the WBCSD's Executive Committee in a key note speech on October 24. It predicts that global energy demand will increase by 60% by 2030, and that 85% of the world's incremental energy needs will be met by fossil fuels. Under this scenario, developing country carbon emissions would double by 2030, surpassing those of the OECD countries. —World Business Council for Sustainable Development

- Today, energy prices are at historic highs. Some analysts estimate that energy price shocks this year could cost American consumers more than $40 billion. Speaking very frankly, we cannot afford this kind of expense. —Jeff Bingaman

- It took us 125 years to use the first trillion barrels of oil. We'll use the next trillion in 30. —Cambridge Energy Research Associates

- In 20 years the world will consume 40% more oil than it does today. —EIA Projections of Oil Production Capacity and Oil Production

- By 2050, at bio-extinction's current rate, between 25 per cent and 50 per cent of all species will have disappeared or be too few in numbers to survive. There'll be a few over-visited parks, the coral reefs will be beaten up, grasslands overgrazed. Vast areas of the tropics that have lost their forests will have the same damn weeds, bushes and scrawny eucalyptus trees so that you don't know if you're in Africa or the Americas. —Stuart L. Pimm

- By 2030 the number of cars in the world will increase by 50%. —Chevron

- By 2050, the world must generate a dollar of GDP with only half the energy used in 2002, equivalent to an economic efficiency improvement of 1.5% per year, a rate of change 20% higher than that achieved in the last 30 years. Additionally, each Tera-Joule of energy used must generate 45% less carbon emissions than in 2002, implying a 1.3% improvement in carbon intensity per year. This is equivalent to twice the global rate of decarbonization in the last 30 years. —World Business Council for Sustainable Development, "Pathways to 2005"

Linking Saving Energy to Other Metrics

- In many businesses, a 20% cut in energy costs represents the same bottom-line benefit as a 5% increase in sales. —Action Energy

- We helped one of our suppliers reduce the packaging on one of their popular toys. As a result, we were able to distribute it using 230 fewer shipping containers, so we saved 356 barrels of oil and 1300 trees. —Wal-Mart Website

- Reducing the energy costs in a large hotel by 10 percent provides the same financial benefit as selling about 930 additional room nights a year. —Energy Star

- For every 10 percent reduction in energy costs, a supermarket can boost profit margins by close to 6 percent. —Energy Star

- For an average school district with six school buildings, a 10 percent energy bill savings can mean the purchase of 30 more computers or 1,500 textbooks. —Energy Star

- Consuming energy derived from fossil fuels contributes directly to the problem of global climate change. Even energy use that seems insignificant really adds up over time. For example, lowering the temperature by 2 degrees in FAS buildings would save $350,000 and prevent 6.6 million pounds of CO_2—the equivalent of 650 cars not driven for a year—from entering the atmosphere. —Harvard Green Campus Initiative

- $75,000 energy savings = 1 new portable X-ray machine, $15,000 = 1 new ER stretcher —Natural Resources Canada and the Canadian College of Health Service Executives

- Through the energy management efforts implemented at comparable mall properties, we reduced electricity usage by 133 million kWhs for 2004 and 2005 combined, as compared to 2003. This represented a 6.8 percent reduction in electricity usage across a portfolio of comparable properties. The EPA estimates this reduction in electricity usage further translates to the avoidance of 84,038 metric tons of carbon dioxide. In addition, the EPA also calculates that this is equivalent to 18,190 cars not driven for one year, 689 acres of forest preserved from deforestation or saved electrical energy to power 10,788 US homes for a full year. —Simon Property Group

- Higher energy costs are here to stay, according to most industry analyst. Since higher oil prices mean goods and people become more expensive to move, the cost of everything from supplies to shipping to sales calls has increased-and, of course, dramatically higher office energy bills can take a serious bite out of the bottom line. How? Consider this example: a company with a 10 percent profit margin must generate $10 in revenue to cover every $1 spent on power for the office. If that company left energy use unchecked and its energy costs increased by $2,000 per month, it would need to bring in an extra $20,000 in revenue to cover the bill. —IBM *Forward View*

Savings at the State Level

- In 2001, California cut energy peak use by 13% in seven months through energy conservation activity. —Flex Your Power

- Energy efficiency results in lower energy bills, which enable consumers to invest their money in other areas. In 2000, U.S. consumers and businesses spent more than $600 billion for total energy use (electricity, natural gas and gasoline). Had the nation not dramatically reduced its energy intensity over the previous 27 years, they would have spent at least $430 billion more on energy purchases in 2000. California saved an estimated $600 million by reducing electrical usage in the first six months of 2001 alone. —California Flex Your Power

- The [California] Department of Consumer Affairs' statewide public information program [Flex Your Power] included coordination with community groups, the Department of Aging, and other state agencies. These efforts appear to have had a wide spread and profound impact on the publics decisions regarding energy use. Voluntary actions by California citizens in both residential and business settings reduced their peak demand by 2,616 megawatts by October 1, 2001. The Department of Consumer affairs and the Energy Commission are conducting research to determine the impact of specific programs on individuals' decisions and how well these effects will endure. —Flex Your Power Summer 2001 Conservation Report

- In the summer of 2001, when called upon, Californian reduced their peak energy use by 5,570 mega watts (MW). 1,000 MW is equivalent to the capacity of one nuclear plant. —California Flex Your Power

Investor Thoughts on Energy Efficiency

- In a paper published last year in the *Financial Analyst Journal,* Jeroen Derwall of the RSM Erasmus University in Rotterdam and colleagues found that the average annual return between1995 and 2003 on a portfolio of companies that ranked high on greenery was 12.2%, compared with 8.9% for low ranked companies. Maybe easier access to capital (which the green investors presumably offer) helps. Maybe it's just that companies that are well managed overall also tend to pay attention to their environmental profile. Either way, greenery seems to go with success. —"A coat of green," *The Economist*

- Organizations that improve energy performance out perform their competitors by as much as 10%—United States Department of Energy

- The environment is one of the most complex challenges facing management, in part because there are high levels of uncertainty as well as many shareholders and complex issues to address. It is implied that companies dealing well with this high level of complexity have the sophistication to succeed in other parts of the business and, thereby, earn superior returns. Energy management is an important aspect of environmental performance which also poses a complex challenge to management. As a result, it is likely that energy management performance is also a strong indication of management quality and stock potential. —Innovest Strategic Value Investors

- Companies with energy-efficient programs are not only more environmentally friendly but also cut costs—Fiona Harvey, *Financial Times*

- Dollars saved are dollars earned. Efficiency initiatives save money that adds to net income and shareholder wealth. Also, plant capacity recaptured through efficiency improvements can support new product lines, in turn providing greater market share and penetration for the manufacturer—thus generating new revenues. —Alliance to Save Energy

Energy and Productivity

- High energy prices have been a primary factor in the loss of 100,000 jobs in the chemical industry alone. —National Association of Manufacturers

- Energy intensive industries that don't pass the increased costs through (e.g. airlines) are left in a weakened, if not fatal, condition. —Alliance to Save Energy

- We currently waste about half of the energy we use in this country, so the potential [for energy savings] is quite enormous. —Kateri Callahan

- If millions of houses cut their consumption by a few percent, the Natural Resources Defense Council argues, then the total demand would fall, and the price would too, saving the economy billions of dollars. —*The New York Times*

- America's CEOs know the challenges that high energy costs pose to our companies and the U.S., and a new Business Roundtable CEO survey shows that energy costs are now among the greatest cost pressures facing companies. —Charles O. Holliday, Jr., Chairman of the Roundtable's Environment, Technology & the Economy Task Force and Chairman and CEO of DuPont

Computer Costs

The exact amount of energy used by data centers, and computers in general, is hard to estimate. However, the consensus is that it is one of the fasted growing end-use segments.

- Recent research from Fujitsu Siemens Computers NV found that 127m pounds ($225m) is wasted every year in the UK alone because employees fail to turn their PCs off when they leave work at the end of the day. —reported by ZD Net

- Turning off computers at the end of each day and utilizing the sleep and hibernation features can save up $45 per computer and $55 per monitor annually. Using an average of one person per 250 square feet and 1.2 computers per person equates to $48,000 a year in a 100,000 square foot building. —reported by DOE

- Energy wasted by computers and monitors costs U.S. organizations, such as hospitals and other healthcare facilities, about $1.5 billion every year. Computers and monitors use more electricity than all other forms office equipment combined. More than half of this energy waste could be prevented if the 60 percent of office computer monitors that are left on at night were turned off, and if the 40 percent of monitors that do not take advantage of their "power management" feature were enabled. —Clark A. Reed, U.S. Environmental Protection Agency

- According to some estimates, just turning off a monitor can save 75 percent of the overall energy consumption of a PC. —Graeme Wearden

- "Sleep" features that power down home office equipment and other electronic devices that are turned on but not in use can save households up to $70 annually. —Alliance to Save Energy, "Power$mart Booklet"

Equipment Costs

On average, a typical office building spends at least 18% of its energy budget powering office equipment.

- If every household in the United States lowered its average heating temperature 6 degrees over a 24-hour period, we would save the equivalent of more than 570,000 barrels of oil per day. —U.S. Navy

- Over an air conditioner's lifetime, only one-fourth of the total cost is for the purchase of the air conditioning unit. The greater cost—three-fourths—is for the energy to run the air conditioner. —Alliance to Save Energy

- Replacing old model air conditioners with Energy Star units can cut cooling bills by 20 percent or more. —Energy Star

- Repairing a seal that leaks a drop of electrically heated hot water every five seconds can save you about 400 gallons of water, 59 kilowatt-hours of electricity, 87 pounds of carbon dioxide, and $7 per year (at 8¢/kWh). —U.S. EPA

- Before you push the "print" button on your photocopier, think. Ask yourself whether you really need so many copies. Perhaps you can get by with a single copy and a routing slip, with electronic mail, or with posting one copy in a high traffic area where your intended reader are sure to see it. Such strategies save both energy and paper. —American Express and Consolidated Edison

- Turn off special purpose machines. Think of the other machines in your office that you could turn off when you are not using them—typewriters, adding machines, mail handlers, paper shredders, room air conditioners, humidifiers, etc. The longer your list, the greater the potential for saving energy. —American Express and Consolidated Edison, NY

- A failed open steam trap with a 3/8-inch orifice at 100-psig pressure loses 4,680,000 pounds of steam annually. —U.S. Navy

- If you're in an office building, you can turn up the thermostat 4 degrees in the summer on hot afternoons. Chances are nobody would even notice, but you could save 20 or 30 percent of that peak energy load. —Amory B. Lovins

Lighting Costs

On average, a typical commercial building will spend about 23% of its energy budget on lighting. Additional energy is needed to offset the heating properties of the lights.

- You can make savings of around 15% of lighting costs just by making people aware of the need for switching off unnecessary lights—Rolls Royce

- If every student in America changed one light bulb to a CFL, it would add up to a savings of more than $2 billion in energy costs. —Kenny Luna

- Substitute compact fluorescent light bulbs for incandescent bulbs. If you replace just four 100-watt incandescent bulbs that burn four or more hours a day with four 23-watt fluorescent bulb, you could save more than $100 over three years. —Energy Secretary Samuel Bodman

- If 300,000 Navy personnel turned off their office lights during the lunch hour (4 fluorescent tubes off per person for 250 hours/year), the Navy could save each year $1.2 million and reduce emissions by 51,600,000 pounds of nitrogen dioxide, 124,800,000 pounds of sulfur dioxide, and 13,656,000,000 pounds of carbon dioxide. —U.S. Navy

- Less than 5 percent of the electricity consumed by an incandescent lamp actually is turned into useful light. Basically, all the electrical energy consumed by the bulb is converted to heat.

- Fluorescent lights convert electricity to visible light up to 5 times more efficiently than incandescent lights and last up to 20 times longer. —U.S. Navy

- If you're in an office with Venetian blinds, tilt them up so that the light is bounced up on the ceiling as God intended. Then the whole room will be suffused with diffuse, soft light, and you'll find that you will see better if you actually turn off the lights. —Amory B. Lovins

Cost of Oil

- U.S. dependence on oil, particularly foreign oil, carries significant economic and political risks. We import 10 million barrels of oil and petroleum products each day—more than half our daily needs. To do so, we send roughly $200,000 each minute overseas to buy oil, contributing significantly to the U.S. trade deficit. Furthermore, the nations dominating the world oil market are located in historically unstable regions of the world, creating complex and delicate relationships for U.S. foreign policy. —Union of Concerned Scientists

- The United States spends over $200,000 every minute to buy foreign oil, accounting for roughly a fourth of our annual trade deficit. Even if we imported far less, our economy would still be susceptible to Persian Gulf politics and OPEC's market power because the price we pay for oil—whether from domestic or foreign supplies—is tied to the world market. Political events in the Middle East precipitated the last three major oil price shocks, and each time the United States experienced an economic recession in their wake. The estimated costs of oil dependence to the US economy are $7 trillion over the past three decades—as much as we paid on the national debt during that period. —Union of Concerned Scientists, Energy Security: *Solutions to Protect America's Power Supply and Reduce Oil Dependence*

- The political instability of the Persian Gulf has caused three major price shocks over the past 30 years. The Iraqi invasion of Kuwait in 1990 took an estimated 4.6 million barrels per day out of the global oil supply for three months. The Iranian revolution reduced global oil supplies by 3.5 million barrels per day for six months in 1979, and the Arab oil embargo eliminated 2.6

million barrels per day for six months in 1973 (EIA, 2001c). In each of these cases, the world oil supply dropped only about 5 percent (Davis, 2001), but world oil prices doubled or tripled (Greene et al., 1998). —Union of Concerned Scientists, Energy Security: *Solutions to Protect America's Power Supply and Reduce Oil Dependence*

GENERAL ENERGY AWARENESS FACTS AND QUOTES

Leadership

- I am personally convinced that one person can be a change catalyst, a "transformer" in any situation, any organization. Such an individual is yeast that can leaven an entire loaf. It requires vision, initiative, patience, respect, persistence, courage, and faith to be a transforming leader. —Stephen R. Covey

- Never doubt that a small group of thoughtful committed citizens can change the world. Indeed it's the only thing that ever has. —Margaret Mead

- You can't always wait for the guys at the top. Every manager at every level in the organization has an opportunity, big or small, to do something. Every manager's got some sphere of autonomy. Don't pass the buck up the line. —Bob Anderson

- I like to things to happen; and if they don't happen, I like to make them happen. —Winston Churchill

- Leaders are visionaries with a poorly developed sense of fear and no concept of the odds against them. They make the impossible happen. —Dr. Robert Jarvik

- It takes a lot more energy to fail than to succeed, since it takes a lot of concentrated energy to hold on to beliefs that don't work. —Jerry Gillies

- A leader takes people where they want to go. A great leader takes people where they don't necessarily want to go but ought to be. —Rosalynn Carter

- It's about changing behaviors… not just light bulbs. —Eugene Matthews, Case Western Reserve University

- I must admit that I personally measure success in terms of the contributions an individual makes to her or his fellow human beings. —Margaret Mead

- Get on with it… just do it. Why wait when there are excellent savings to be realized? —Brain Manser, Building Service Manager

- Energy is a solo performance that only you can control.

- There is nothing wrong with America that the faith, love of freedom, intelligence and energy of her citizens cannot cure. —Dwight D. Eisenhower

Little Actions by Individuals can Make a Big Difference

- Reducing electricity consumption by as little as 50 kWh per day can save as much as $1,500 to $2,000 per year. —Peninsulas Health Care Corporation

- The ocean is made of drops. —Mother Teresa

- Employee or tenant behavior can have substantial impacts on building energy use as they influence the power required for lighting, computer operation, and heating, among other energy uses. Promoting energy awareness among staff can provide positive returns quickly for a small up-front cost. —Energy Star

- The important thing is not to ignore the ideas that bring small savings—they grow together into larger ones. —Brian Docherty, Maintenance Controller

- A single degree of over-heating or over-cooling on campus costs UB $100,000 a year. —University of Buffalo Energy Awareness Website

- By taking smart and responsible actions… we can be successful at minimizing the pain caused by high energy costs this winter. —Energy Secretary Samuel Bodman

- Consuming energy derived from fossil fuels directly contributes to the problem of global climate change. Even energy use that seems insignificant really adds up over time: for instance, our research shows that one desktop computer and 17″ CRT monitor left on 24 hours per day for one year can release over 1500 pounds of carbon dioxide. The manufacturing, distribution and disposal of computers, as well as the air conditioning often necessitated by their operation, release even more greenhouse gases. In addition, these activities consume natural resources and release toxic chemicals into the environment. The positive thing about these figures is that small changes will have major impacts. With 13,000 computers in FAS, even a slight increase in turn-off rates will prevent tons of CO_2 from entering the atmosphere. Just think of how much we could save with a really concerted effort! —Harvard Green Campus Initiative

- Tenants directly impact over 50% of the energy consumed in a typical office building. —DOE presentation

- If just one in ten homes used Energy Star® qualified appliances, the environmental benefit would be like planting 1.7 million new acres of trees. —Energy Star Program

- Remember, if you can cut your emissions by 70 per cent, so can the business you work for. By doing so, in the medium term the business will save both money and the environment. And because society so desperately needs advocates—people who can act and serve as witnesses to what can be done and should be done—by taking such public actions you will be achieving results way beyond their local impact. —Tim Flannery, *The Weather Makers*

- If everyone who has the means to do so takes concerted action to rid atmospheric carbon emissions from their lives, I believe we can stabilize and then save the cryosphere—those parts of the world, such as the poles, where water is frozen. Tim Flannery, *The Weather Makers*

- American households typically spend more than $200 annually on air conditioning. Households in some regions of the South can easily spend twice that much. —Alliance to Save Energy

- The U.S. consumes a million dollars worth of energy every minute. –U.S. Energy Information Administration

- You use 25 barrels of oil a year. —Energy Information Administration

- Almost 50% of California's peak energy demand is residential and commercial air conditioning and commercial lighting. —California Flex Your Power

- On average, UCSD's (University of California San Diego) electrical consumption costs the campus $40,000 per day. You can help save energy and our environment by the following steps.
  - — Turn off lights, and computers and appliances when not in use.
  - — Use laptops and inkjet printers which require 90% less energy than desktop computers and laser printers.
  - — Close widow shades to keep rooms cool.
  - — Set thermostats to 78° on hot days and 68 on cold days.
  - — Dress comfortably for the weather.
  - — Do laundry after 7 p.m.

  —Environmental Essentials, UCSD's green Guide

- Our daily decisions, be it the setting on our thermostat, the choice between using a car or public transport, or even choosing a long-haul holiday destination against a regional one, influence energy use somewhere along the value chain. A shift in consumer choices directly or indirectly affects the other megatrends due to the fact that it is consumption that ultimately drives economic activity. Since many small decisions can add up to make a tremendous difference, a megatrend shift in our consumption choices (lifestyle changes) can make an important contribution to a carbon-constrained world. —World Business Council for Sustainable Development, *Pathways to 2050*

- A journey of a thousand miles begins with a single step. —Modified Chinese proverb

Tips on Motivating Others

- What you always do before you make a decision is consult. The best public policy is made when you are listening to people who are going to be impacted. Then, once policy is determined, you call on them to help you sell it. —Elizabeth Dole

- People seldom improve when they have no other model but themselves to copy. —Oliver Goldsmith

- In the arena of human life the honors and rewards fall to those who show their good qualities in action. —Aristotle

- People never improve unless they look to some standard or example higher and better than themselves. —Tyron Edwards

- I multiplied myself by my activity. —Napolean

- Activity is contagious. —Ralph Waldo Emerson

- Example has more followers than reason. —Christian Bovee

- I start with the premise that the function of leadership is to produce more leaders, not more followers. —Ralph Nader

- True vision is always two fold. It involves emotional comprehensions as well as physical perception. —Ross Parmenter
- Knowledge is power, but enthusiasm pulls the switch. —Ivern Ball
- No man will work for interests unless they are his. —David Seabury
- Put your trust in people. —Churchill
- When there is no vision people perish. —Ralph Waldo Emerson
- Life is the sum of all your choices. —Albert Camus
- He who dares... Wins!—British SAS (Special Air Service) motto
- You'll get more stares, if you take the stairs! —Sign at an outdoor escalator at the San Diego Airport
- He that will not apply new remedies must expect new evils. —Francis Bacon
- Life is always at some turning point. —Ovid
- As we learn we always change, and so our perception. This changed perception then becomes a new teacher inside each of us. —Hyemeyohsts Storm
- Ninety percent of the game is half mental. —Yogi Berra
- The time is always right to do what is right. —Martin Luther King Jr.
- Creativity involves breaking out of established patterns in order to look at things in a different way. —Edward de Bono (19334) Maltese Physician, Educator

The Importance of Team Work

- The achievements of an organization are the results of the combined effort of each individual. —Vince Lombardi

- An empowered organization is one in which individuals have the knowledge, skill, desire, and opportunity to personally succeed in a way that leads to collective organizational success. —Stephen R. Covey

- To succeed as a team is to hold all of the members accountable of their expertise. —Mitchell Caplan

- Great organizations demand a high level of commitment by the people involved. —Bill Gates

Executive Involvement

- Being a leader means being accountable. To your company. To your shareholders. To your environment. —U.S. EPA

- Leaders would be more respected, not less, if they delivered the bad news. —Marshall Goldsmith

- Make no little plans; they have no magic to stir men's blood and probably will themselves not be realized. Make big plans; aim high in hope and work.. —Daniel H. Burham

- The success lies in its credibility in the eyes of staff and other members of management. In the case of Brandon Regional Health Authority there was executive level buy-in from the beginning making a high profile committee within the institution: lending both public and private support for the initiatives and affording member a sense of ownership and accomplishment. From a financial perspective, it was crucial that management be supportive otherwise funding for the many initiatives could have halted any possible progress. —Gordon Neal, Chief Power Engineer

- The leaders who work most effectively, it seems to me, never say "I." And that's not because they have trained themselves not to say "I." They don't think "I." They think "we"; they think "team." They understand their job to be to make the team function. They accept responsibility and don't sidestep it, but "we" gets the credit.... This is what creates trust, what enables you to get the task done. —Peter Ducker

- The only safe ship in a storm is leadership. —Faye Wattleton

The Importance of Communicating and Educating

- In spite of soaring fuel and electricity prices many companies remain ignorant of the energy efficiency message. —*Financial Times*

- Merely asking questions about energy consumption or patterns of use has a strong motivating effect. When people realize that someone is paying attention they respond. This alone can reduce consumption by 3 -5%—If you don't believe this try it yourself! —Carbon Trust

- The "Shut it off" program, educating employees to do the obvious but easily overlooked, helped focus attention on the energy reduction goal. —United Technologies Corporation

- An investment in knowledge pays the best interest. —Benjamin Franklin

- The conservationist's most important task, if we are to save the earth, is to educate. —Peter Scott

- Knowledge is power. —Sir Francis Bacon

- USAA Real Estate Company included an aggressive communications plan to involve their tenants. In addition to promoting energy-efficient behavior such as powering down idle equipment, effective communication led USAA to realize that few tenants were in the office on Saturdays. This enabled them to turn back their heating and cooling equipment on the weekends. —EPA

- Our ignorance is not so vast as our failure to use what we know. —M. King Hubbert, Geophysicist

- Vision is the art of seeing things invisible. —Jonathan Swift

- Thought is the seed of action. —Ralph Waldo Emerson

- Only the educated are free. —Epictetus

Conservation Attitudes

- Energy conservation is more than a private virtue; it's a public virtue. —George W. Bush

- Personally, I feel the time has come to act—to take steps as a nation to reduce the carbon intensity of our economy. And it's going to take all of us to do it. —Paul Anderson, CEO Duke Energy

- Energy is often regarded as an unseen and uncontrollable overhead with no one taking responsibility. —Action Energy

- I used to be skeptical of global warming, but now I'm absolutely convinced that the world is spiraling out of control. CO_2 is like a bushfire that gets bigger and bigger every year. All of us who are in a position to do something about it must do something about it. —Richard Branson, Founder of the Virgin Group

- The good news is that what makes economic sense often also makes environmental sense. —George Caraghiaur

- For many people, the idea of energy conservation and energy efficiency is just a noble ideal, a commitment that often wanes, something non-essential to their lifestyle. For low-income households, conservation and efficiency is a way of life; it is a means to sustain their lifestyle, it is a way to gain more control over their household expenses, thus control over more aspects of their lives. —Jim Morton

- The most important environmental issue is one that is rarely mentioned, and that is the lack of a conservation ethic in our culture. —Gaylord Nelson

- The ultimate test of man's conscience may be his willingness to sacrifice something today for future generations whose words of thanks will not be heard. —Gaylord Nelson

- We do not inherit the earth from our ancestors, we borrow it from our children. —Native American Proverb

- I want to be able to tell my grandson when I'm in my later years that I didn't turn away from the evidence that showed we were doing some serious harm. —Al Gore

- What's near and dear to my heart is cooperative conservation. —Gale A. Norton

- One thing is clear: the decisions we make now will determine whether we leave to future generations a healthy, livable world that holds the same promise of opportunity as the one that we inherited from the past. —Al Gore

- It seems to me that we all look at Nature too much, and live with her too little. —Oscar Wilde

- We've known for some time that we have to worry about the impacts of climate change on our children's and grandchildren's generations. But we now have to worry about ourselves as well. —Margaret Beckett, British Secretary of State for the Environment

- An overwhelming majority of consumers—92 percent—agree that business, government, and consumers have an equal responsibility to reduce energy use. —Alliance to Save Energy, 2003 Consumer Market Research

- ...I think people are pretty smart, and they have plenty of incentive to use energy in a way that saves money. Some of what we need to do in the short term is curtailment, which needn't be painful, it's usually just turning off things you're not using anyway. The off switch is the best way to cut your bill this summer. —Amory B. Lovins

- It's not that Americans have anything against saving energy. It's just that investing in energy efficient technologies often require an extra effort for them—whether it's the additional knowledge they need to gain to make an informed decision or the additional investment they may have to make to purchase the technologies. —Michael Evans, EPRI.

- Eighty percent of American consumers agree that America needs to reduce oil imports. —Alliance to Save Energy, 2003 Consumer Market Research

- Change our electric bulbs, our boilers and our refrigerators, insulate our houses, buy less polluting cars, use public transport: these are some of the things we should do if we want to protect the environment and guarantee a stable supply of energy for our children. —European Commission Directorate General for Energy and Transport, "20% energy savings by 2020," a green paper on energy efficiency

- True conservation provides for wise use by the general public. The American people do not want our resources preserved for the exclusive use of the wealthy. These land and water resources belong to the people, and people of all income levels should have easy access to them. —George D. Aiken

- The rich resources of our planet Earth have given birth to many forms of life and have supported the wide-ranging and ambitious activities of mankind. Nevertheless, recent activities have exceeded the life-sustaining abilities of the Earth. This poses a threat not only to our coexistence with other forms of life on this planet, but also to the future of the human race itself. Global environmental conservation is the most urgent issue that the whole of mankind faces. We must be committed to restoring the Earth to its full capacity, and pass this on to future generations. To achieve this, we need to be more aware of the importance of the Earth in our personal lives, not just in our businesses, and strive to continuously modify our corporate activities and lifestyles to reduce the impact our society as a whole has on the global environment to a level that the planet can cope with. —Masamitsu Sakurai, Chairman of the Board, President and Chief Executive Officer of Ricoh Company

- Energy efficiency helps us reach our key corporate goals by improving financial returns, effecting positive social and environmental change in our communities, and being part of the solution to climate change. —Dave Mowat, CEO of Vancity

- ...the results suggest that individual control over appropriately designed lighting can result in energy savings. The lit environments people selected for themselves had, on average, lower power requirements compared with environments in line with recommendations in existing codes and standards. In other words, giving people control over lighting might result in lower energy consumption compared with a fixed lighting design with lighting power density at the maximum allowed by codes and standards. —Jennifer A. Vetich, and Guy R. Newsham, *Energy Control can be Energy Efficient*

- The Earth's well-being is ...an issue important to America—and it's an issue that should be important to every nation and in every part of the world. My Administration is committed to a leadership role on the issue of climate change. We recognize our responsibility, and we will meet it—at home, in our hemisphere, and in the world. —George W. Bush

- [Climate change is] the only thing that I believe has the power to fundamentally end the march of civilization as we know it, and make a lot of the other efforts that we're making irrelevant and impossible. —Bill Clinton

- Conservation benefits the consumer not just by lowering bills, but by providing a long-term resource. Conservation allows us to manage risk, stabilize our energy system, and at the same time, fulfill our stewardship obligation to the region. —Joan Smith, Oregon Public Utility Commissioner

- Energy conservation is the cornerstone of a clean and affordable energy future. Since 1981, Northwest utilities have saved enough energy through conservation to power a city one and a half times the size of Seattle. Energy conservation protects our air and water while keeping our homes and businesses comfortable at a lower cost than natural gas or coal-powered generation. —Ralph Cavanagh, NW Energy Program Director, Natural Resources Defense Council

- We must have a relentless commitment to producing a meaningful, comprehensive energy package aimed at conservation, alleviating the burden of energy prices on consumers, decreasing our country's dependency on foreign oil, and increasing electricity grid reliability. —Paul Gillmor

- I think there are few more challenging and worthwhile jobs in the world today than meeting the energy needs of a developing world in a sustainable way; few more stimulating than using technology and management innovation to solve fundamental problems-like tackling climate change-where creativity is embraced and applied. —Sir Mark Moody-Stuart

- Climate policy has been held hostage to a tacit presumption that if saving a lot more energy were possible at an affordable price, it would already have been implemented. That's like not picking up a $100 bill from the sidewalk because if it were real, someone would previously have picked it up; or like an entrepreneur who abandons a good business idea because if it were sound, it would have been done earlier. —Amory B. Lovins

Addiction and Energy Dependence

- When we look across all energy markets, we see an alarming picture. In past energy crunches, U.S. consumers could switch to another fuel when supplies were tight. Today demand for petroleum fuels, natural gas, and coal is surging beyond the markets' ability to deliver. High coal and natural gas prices are in turn driving electricity prices higher in many regions. Dr. Neal Elliott, ACEEE's Industrial Program Director

- America is addicted to oil. —President George Bush, in a 2006 State of the Union address

- Energy consumption rose hand in hand with economic growth during industrialization. This cannot go on forever. Energy, after all, is a limited resource while economic growth can, theoretically, continue indefinitely. —Ernst von Weizsäcker, "Less energy. More wealth"

- When the well's dry, we know the worth of the water. —Benjamin Franklin

- Since the beginning of our short oil era around 1860, world population has increased dramatically. This population growth has been fueled substantially by oil. In the United States, food travels more than 1,000 miles on average, requiring over 10 times the petroleum energy to produce than its solar energy food value (calories). As a practical matter, we are eating mostly petroleum. —Francis de Winte and Ronald B. Swensen, "Dawn of the Solar Era, A wake-up call"

- With only 2 percent of the world's proven oil reserves but 26 percent of the world's consumption, the United States cannot significantly reduce the costs of oil dependence by drilling at home. Saving oil is the most secure strategy we can take to insulate the US economy from price shocks, counteract the market power of OPEC, and reduce the amount of money we send abroad. —Union of Concerned Scientists, "Energy Security: Solutions to Protect America's Power Supply and Reduce Oil Dependence"

- US energy production is highly centralized, making it vulnerable to well-placed attacks. A typical nuclear power plant serves 1.25 million homes, a large coal power plant half a million homes. Concentration in the oil business is even greater, with only 150 refineries nationwide producing fuel for over 200 million cars (EIA, 2001f). A major disruption at any one of these energy supply facilities can endanger lives and affect prices for hundreds of thousands of customers. —Union of Concerned Scientists, "Energy Security: Solutions to Protect America's Power Supply and Reduce Oil Dependence"

- This year alone, China expects to add a staggering 81 gigawatts of new capacity. Over the next five years, the government plans to invest 600 billion yuan ($75 billion) in still more power plants. The state Grid Corporation of China, meanwhile plans to spend 800 billion yuan over the same period, expanding and upgrading its transmission networks. —"Power to the people," *The Economist*, February 11th, 2006

- A disruption at a key power plant, refinery, transmission hub, or pipeline can break the flow of power or fuel to millions of customers and create costly energy price spikes. A major accident at a nuclear power plant could kill tens of thousands and contaminate an area the size of Pennsylvania. Mock intruder tests at US nuclear facilities have shown that reactor damage could result from attacks far less threatening than those of September 11. —Union of Concerned Scientists, "Energy

Security: Solutions to Protect America's Power Supply and Reduce Oil Dependence"

- Energy efficiency helps ensure that California has stable and reliable electricity service by making the system less vulnerable to electricity supply shortages. California relies on natural gas and hydroelectric generation plants (primarily in the Pacific Northwest) for more than half of its electricity. Neither source is very reliable. The entire nation currently is experiencing a natural gas shortage. During 2000 and 2001, historic drought conditions in the Pacific Northwest decreased the amount of hydroelectric power produced in the Sierras by 25% to 35%. Energy efficiency improvements reduce overall demand and, more important for reliability, shave demand at peak hours —those hours when energy demand is highest and blackouts are likely to occur if supply is low. —Flex Your Power

- Energy efficiency also enhances electric system reliability by reducing the amount of electricity that has to be transported across congested transmission and distribution lines. —Flex Your Power

- In 1994, for the first time in its history, the United States imported more than 50 percent of its petroleum, a level of dependence that aggravates the trade deficit and leaves the American economy vulnerable to oil price shocks. —U.S. Navy

- The more energy-efficient we become as a nation, the less we need to develop additional energy sources. —Lamar S. Smith

- Energy conservation is the foundation of energy independence. —Thomas H. Allen

Energy Management Requires Cooperation to be Successful

- Although technology may postpone the date, we will inevitably face a transition from the present pattern of expanding extraction and consumption to one of careful management of a steady and finite supply of energy. This transition implies widespread changes in people's behavior, involving increased energy conservation and a shift to dependence on renewable sources of energy. —Paul Sern and Eileen Kirkpatrick

- We believe that annual energy bills and greenhouse gas emissions can be significantly reduced through raising building user awareness. Raising awareness will lead to behavioral changes in people and many constructive suggestions for projects. —Northern Territory Department of Planning and Infrastructure, Australia

- You've probably heard people talk about conservation. Well, conservation isn't just the business of a few people. It's a matter that concerns all of us. It's a science whose principles are written in the oldest code in the world, the laws of nature. The natural resources of our vast continent are not inexhaustible. But if we use our riches wisely, if we protect our wildlife and preserve our lakes and streams, these things will last us for generations to come—Walt Disney

- The neoclassical economy predicts that energy savings which can pay will be undertaken, and that energy savings which do not pay will not be carried through. However, the empirical investigations suggest that situations where energy savings are not immediately undertaken also exist, although they are profitable, and that there are situations where energy savings are carried through although they are not profitable. In these situations the neoclassical economy seems to lack explanatory power. —Lene Holm Pederson

- Energy conservation is a local issue, and in a retail operation with more than 350 stores, this means creating awareness and buy-in by local managers. These managers tend to be non-technical and are primarily concerned with merchandising. Zellers, through its 'Energy Road Show' and incentive program, has been able to make energy conservation an issue of continuous concern. —Fred Ware, C.E.T., Senior Manager, Energy, Environment, Sourcing Initiatives, Store Operations, Zellers Inc.

Indecision

- Whenever we seek to avoid the responsibility for our own behavior, we do so by attempting to give that responsibility to some other individual or organization or entity. But this means we then give away our power to that entity. -M. Scott Peck

- A person may cause evil to others not only by his actions but by his inaction, and in either case he is justly accountable to them for injury. —John Stuart Mill

- The future is not in the hands of fate but in ours. —Ambassador Jean Jusserand

- You must constantly change and adapt to a new environment. —Jong-Yong Yun

Over Zealousness

- Leaders shouldn't attach moral significance to their ideas: Do that, and you can't compromise. —Peter F. Drucker

- Voluntary action tends to be more economically efficient than imposed solutions. —Dean Olmstead

- The key to successful leadership today is influence, not authority. —Kenneth Blanchard

- You do not lead by hitting people over the head—that's assault, not leadership. —Dwight D. Eisenhower

- Don't confuse motion with action. —Ben Franklin

- Haste in every business brings failure. —Herodotus

- The map is not the territory. —Alfred Korzybsk

- Keep away from people who try to belittle your ambitions. Small people always do that, but the really great make you feel that you, too, can become great. —Mark Twain

Persistence

- The awareness message is not a one time thing. It must be delivered on a continuing basis. —Gloria Zive, VP Planning & Support

- Our energy campaign is an ongoing project ...we are constantly adding elements to our energy awareness project as we move ahead to keep the message fresh and interesting. Tom Meevis, Environmental Affairs Coordinator

- Accomplishment and success are often the result of commitment and perseverance rather than skill or talent. —George Van Valkenburg

Planning, Measurement and Feedback

- Immediate feedback is the key to successful energy management. –John Dilliott

- To effectively manage energy systems, some level of metering is necessary. Success lies in making use of the metering information. —National Renewable Energy Laboratory

- Computerized monitoring and control equipment allows management to track energy use throughout the company and benchmark performance. This capability provides vital feedback on the success or failure of various [energy] initiatives and alerts management to problems, such as unusual spikes in usage. Monitoring is a key element in the deployment of energy management plans since it measures improvements and educates management about costs and savings. —Innovest Strategic Value Investors

- Provide relevant energy facts such as $75,000 energy savings = 1 new portable X-ray machine, $15,000 = 1 new ER stretcher—Natural resources Canada and the Canadian College of Health Service Executives

- You cannot control what you cannot measure. Measured data is of little use without qualifies analysis. It takes action to get results. —Thomas Mort, Energy Manager

- On examining its energy consumption patterns a small company found that a production process was switched on each night just to keep the night watchman warm. —Action Energy

- Advanced meters can capture electrical anomalies such as transients, voltage disturbances, power factors, and harmonics to troubleshoot poor quality problems. This can be especially useful when monitoring sensitive loads. Transients cause printed circuit boards in computers and other electronic equipment to fail prematurely. Improper power factors can result in surcharges from utility companies. High harmonics can shorten the life of transformers. Using advanced meters allows power quality problems to be detected and documents so solutions can be developed and implemented. —National Renewable Energy Laboratory a U.S department of Energy Laboratory, *Advanced Utility Metering*

- Goals are not only absolutely necessary to motivate us. They are essential to really keep us alive. —Robert H. Schuller

- Whenever an individual or a business decides that success has been attained, progress stops. —Thomas Watson

- To accomplish great things, we must not only act, but also dream; not only plan, but also believe. —Anatole France

- If you don't know where you are going …You might end up somewhere else. —Yogi Berra

Rewarding Energy Efficient Behavior

- Awareness campaigns benefit greatly from the presence of small tokens of participation such as pens, Post-it note pads, magnets, buttons, key rings, mini-screwdrivers, badge holder, t-shirts & caps, mugs, etc. —Natural resources Canada and the Canadian College of Health Service Executives

- Food Lion's energy awareness plan rewards maintenance staff by awarding quarterly bonuses for improving energy performance. Keeping maintenance staff motivated to save energy has helped Food Lion reduce its utility cost per store per week by 5.5 percent. —Energy Star, *Guidelines for Energy Management*

- Hilton Hotels tied hotel general managers' annual bonuses to energy performance. By meeting Hilton's goal of reducing energy consumption by 5 percent for every owned hotel, a manager's annual bonus was increased. This resulted in almost every property meeting the goal. —Energy Star, *Guidelines for Energy Management*

Specific Devices and Energy Use

- In the U.S., copiers use about 7 TWh/year of electricity, and a similar amount of energy is embodied in the estimated 2.2 million tons/year of paper used in copiers. These cost the economy about $500 million/year for the electricity and $2.2 billion/year for paper. Bruce Nordman, Mary Ann Piette, Brian Pon, and Kristopher Kinney, "It's Midnight ... Is Your Copier On?: ENERGY STAR Copier Performance"

- Computers in general use over 1% of the nation's commercial electricity and more energy than all office equipment combined. A monitor uses roughly 60 to 90 watts when active but only 2 to 5 watts when in a low-power sleep mode. 95% of monitors have the ability to go into this low-power sleep mode automatically after a period of inactivity. In general, an organization can annually save about 200 kWh per computer. For an organization with 1,000 computers, that's 200,000 kWh savings per year!—Energy Star, *How Utilities and REPs Use Monitor Power Management.*

- After contacting ENERGY STAR, the [Cisco] company found a way to activate power management features on 20,000 monitors in its California offices using only 16 hours of IT time. With power management, Cisco expects to save about $528,000 dollars annually, and because the company has reduced its energy consumption by more than 3.4 million kilowatt hours, it is also helping reduce air pollution. ... Building on the successes of its California offices, the company is expanding the program to 30,000 monitors at other locations. All told, the combined savings from the "sleeping" monitors will be 8.5 million kilowatt hours—enough energy to power about 10,000 U.S. households for one month—and about $1 million a year. Cisco is among the first companies in the nation to take advantage of the "Sleep is Good" campaign, and there are many more kilowatts and dollars to be saved. In fact, businesses and organizations across the nation can save a total of more than $900 million and 11 billion kilowatt-hours with monitor power management. What's your company's share? —"More Sleep Will Increase Cisco's Bottom Line by $1 Million a Year," Steve Ryan, EPA Program Manager

- Lighting accounts for around 20% of the electrical energy used in the United States. In commercial buildings, lighting accounts for about 25—30% of energy consumed and is a primary source of heat gain and waste heat in the buildings. —Susannah Patton, "Powering Down," *CIO*

- A typical 10,000 square-foot data center consumes enough juice to turn on more than 8,000 60-watt light bulbs. That amount of electricity is six to 10 times the power needed to operate a typical office building at peak demand, according to scientists at Lawrence Berkley national laboratory. Given that most data centers run 24/7, the companies that own them could end up paying millions of dollars this year just to keep their computers turned on. —Susannah Patton, "Powering Down," *CIO*

- It is good to provide an occasional breakout of the estimated energy cost by source device. Individuals tend to under or over estimate the contribution of individual devices. A good example comes from the University of California San Diego:

In 2003, UCSD consumed 600,000 kilowatts of electricity daily costing the campus $39,000 per day. In 2004, we have more students and more space, yet we need to use LESS electricity than we did in 2003. —John Dilliott, UCSD

Average annual energy costs for common plug-in items [for UCSD]:

Space heater	$43
Aquarium	$65
Fountain	$19
Microwave	$47
Desktop computer with traditional monitor	$136 @ 24/7; $90 @ 16/7
TV with DVD player	$78
Stereo	$42
Video games	$234
Hair dryer	$62
Housing refrigerator	$101
Laptop w/flat screen monitor	$68 @ 24/7; $45 @ 16/7

User Behavior and Turn-off Rates

- Of the 1,280 computers audited (desktop and desksides only; does not include servers or portables) 44 percent were off, 3 percent were in low-power mode, and the remaining 54 percent were on (these numbers do not total 100 percent because of rounding). —Carrie A. Webber, Judy A. Roberson, Richard E. Brown, Christopher T. Payne, Bruce Nordman, and Jonathan G. Koomey, "Field Surveys of Office Equipment Operating Patterns"

- Through night-time audits, we estimated the fraction of computers, monitors and printers left operating after business hours. The fraction of computers left on at night was found to average near 30%, and range as high as 70%. Rates near 20 to 25%

were found where some prior user education had been done. Persistent in-house conservation advocates were able to achieve a 10 to 15% rate despite some resistance from misinformed IS staff. An office with a mandatory nightly server logoff for data backup produced a similar result. An office with a program of individual training, including a mandate to shut down nightly, produced about a 5% leave-on rate; a similar rate was observed in an office with a high level of environmental awareness. Roger E. Picklum, Bruce Nordman, Barbara Kresch, "Guide to Reducing Energy Use in Office Equipment"

- ...We sent email to people in the target departments, and placed posters on bulletin boards. The messages we used emphasized the pollution and monetary benefits of reduced electricity use, and stressed turning off equipment at night. Behavior change is significant in all cases, and is most pronounced where initial behavior was to leave the most computers operating at night.

  While we can't predict the lowest level of leave-on rate achievable, we did observe levels under 10% for personal equipment in an office with high environmental awareness, and under 5% in an office which had performed hands-on user training. A persistent program of user awareness may be able to achieve leave-on rates near these levels in most offices, and maintain it over time. Offices which require PCs to be on at night can still turn off monitors, and are prime candidates for PC power management. Roger E. Picklum, Bruce Nordman, Barbara Kresch, "Guide to Reducing Energy Use in Office Equipment"

Developing Countries

About one quarter of the world population has no access to electricity. As countries develop so does their appetite for energy. In 2005, China built one large coal-fired electrical power nearly every two weeks. In 2002, total energy use per capita was 4,878 kg-oil-equivalent in developed countries versus only 869 kg-oil-equivalent in less developed counties.

- The biggest single challenge for this century: energy is needed for 10 billion people worldwide and current technology will not provide it in a clean and sustainable way. —International Energy Agency

- When it comes to energy and development, it's easy to get lost in the numbers: millions here, billions there, and trillions into the future. Consider these four: 2 billion people without modern energy services; 500 billion dollars invested annually in energy infrastructure; and 4 billion tons of CO_2 dumped into the atmosphere every year from a 60 trillion dollar global economy. Taken together, they add up to a daunting challenge to the world. —Monique Barbut, "The energy numbers game"

- If current trends continue, china will surpass the U.S. as the world's largest emitter of greenhouse gases in the next decade. —Barbara Finamore

- Oil products demand in the transport sector will grow very fast due to increasing affluence and increasing demand for mobility. This demand would be particularly robust in Eastern and Central European countries, Russia, Asia, and Latin America. —International Energy Agency, "ENERGY TO 2050 Scenarios for a Sustainable Future"

- China was the world's second largest consumer of petroleum products in 2004, having surpassed Japan for the first time in 2003, with total demand of 6.5 million barrels per day (bbl/d). China's oil demand is projected by EIA to reach 14.2 million bbl/d by 2025, with net imports of 10.9 million bbl/d. As the source of around 40% of world oil demand growth over the past four years, with year-on-year growth of 1.0 million bbl/d in 2004, Chinese oil demand is a key factor in world oil markets. —U.S DOE Energy Information Administration

- Developing countries at the beginning would try to catch-up with the same transport demand patterns and behavioral models of industrialized ones. In most cases the construction of an important railroad infrastructure (as done by rich countries in the early days of their industrialization) would not be considered interesting, and priority would be given to road infrastructure. —International Energy Agency, "ENERGY TO 2050 Scenarios for a Sustainable Future"

- If the planet's 6.2 billon non-Americans used as much air conditioning per person as we do, the world's total electrical requirement for residential air conditioning per person would annually come to 4 trillion kilowatt hours. That exceeds the combined electricity supply of China, India, Indonesia, Brazil, Bangladesh, Pakistan, Russia, Nigeria, and Mexico—which together serve as home to more than half of humanity. —Stan Cox, "Cooling Malls Heats the Planet"

- The U.S. Environmental Protection Agency estimates that on certain days almost 25 percent of the particulate matter in the sky's above Los Angeles can be traced to China. Some experts predict that China could one day account for a third of California's air pollution. —Terence Chea, "Chinese industrial growth brings pollution to West Coast"

- And as we work to solve our energy dependence—dependency, we've got to remember that the market for energy is global and America is not the only large consumer of hydrocarbons. As the economies of nations like India and China grow rapidly, their demand for energy is growing rapidly, as well. It's in our interest to help these expanding energy users become more efficient, less dependent on hydrocarbons. You see, by helping them achieve these goals, it will take pressure off the global supply and it will help take pressure off price for American consumers. —George W. Bush

- In 2000, only one in six of us on this planet had access to the energy required to provide us with the high living standards enjoyed in developed countries. Yet these one billion people consumed over 50% of the world's energy supply. By contrast, the one billion poorest people used only 4%. None of us finds poverty acceptable, so the world has set itself various goals to eradicate poverty and raise living standards. These goals require energy, the driver of modern living standards. Increased access to modern energy services such as electricity is a decisive factor in escaping the poverty trap; it vastly enhances opportunities for industrial development and improves health and education. —World Business Council for Sustainable Development, "Facts and trends to 2050, Energy and climate change"

- The *International Energy Outlook 2005 (IEO2005)* reference case projects that world net electricity consumption will nearly double over the next two decades. Over the forecast period, world electricity demand is projected to grow at an average rate of 2.6 percent per year, from 14,275 billion kilowatt-hours in 2002 to 21,400 billion kilowatt-hours in 2015 and 26,018 billion kilowatt-hours in 2025 (Figure 58). More than one-half (59 percent) of the projected growth in demand occurs in the emerging economies, with the mature market and transitional economies accounting for 28 percent and 14 percent, respectively —Energy Information Administration, "International Energy Outlook 2005"

- The business-as-usual-perspective on global energy demand is neither sustainable nor acceptable, Mr. Mandil of IEA told the WBCSD's Executive Committee in a key note speech on October 24 [2005]. It predicts that global energy demand will increase by 60% by 2030, and that 85% of the world's incremental energy needs will be met by fossil fuels. Under this scenario, developing country carbon emissions would double by 2030, surpassing those of the OECD countries. —World Business Council for Sustainable Development

- The world's most populous and fastest-growing nation is eating up a growing share of the planet's oil and coal, pushing up international energy prices and increasingly being forced to look beyond its borders for supplies. In the next 15 years, demand is expected to double, which would bring about a change in the balance of power—not just in terms of electrical supply, but diplomacy, security and finance. —Jonathan Watts, "China's growth flickers to a halt"

- India's **use of oil** has doubled since 1992, while China went from near self-sufficiency in the mid-1990s to the world's second largest oil importer in 2004. Chinese and Indian oil companies are now seeking oil in countries such as Sudan and Venezuela—and both have just started to build what are slated to be two of the largest automobile industries in the world. —World Watch Institute

Energy Efficiency

- The world needs to recognize that the absolute best source of energy is conservation and efficiency. It is extremely unlikely that we will suddenly discover vast supplies of oil or natural gas sufficient to satisfy the world's needs. Technologies such as fuel cells are early in the development phases and have yet to be proven as economic alternatives. We can and must be more efficient in a thousand different and proven ways today. Witness the explosion of Hybrid autos at this year's 2006 international auto show and you see a ray of hope. Solar needs to continue to be developed and installed prices are coming down as manufacturing technology continues to improve. We need to tirelessly work to make every appliance, every car, every home, every business, and every industry as efficient as possible. We need to be as Green as we can be. —Timothy B. Janos, 2006 President of the Association of Energy Engineers

- Addressing energy efficiency is a very proactive approach to take particularly in light of ever-increasing utility cost. It helps us get the most out of our funding. —Mark Nore, Director of Support Services

- Energy efficiency is the cornerstone of sustainable energy policy... Our continuing straitjacket situation shows that without moderating energy demand, no supply strategy will be able to keep up. And with persistent market barriers hobbling energy efficiency investment, we need much bolder efficiency policies to meet these unprecedented energy challenges. —Bill Prindle, ACEEE's Deputy Director

- Since 1973 alone, improvements in energy efficiency have resulted in a 50% reduction of our daily energy intensity per gross domestic production unit, which is the same as discovering 25 extra million barrels of oil equivalent every single day. —American Council for an Energy-Efficient Economy

- By 2050, global carbon emissions would need to be at levels similar to 2000, but also trending downward, in contrast to a sharply rising demand for energy over the same period. No single solution will deliver this change, rather we need a mix of options which focus on using energy more efficiently and lowering its carbon intensity. Changes in supply and demand can help us shift to a truly sustainable energy path. While change takes time, starting the process now and laying foundations for the future are matters of urgency, and business has a key role to play. —World Business Council for Sustainable Development, "Facts and trends to 2050, Energy and climate change"

- Energy efficiency has gained a reputation as one of the 'softer' sciences of cost-reduction and continuous improvement but, done properly, it is clear that it can really pack a punch. —Debbie Giggle

- Maximizing Energy Efficiency and Renewable Energy IS the domestic epicenter in the War on Terror and it is imperative that we maximize the partnerships between the public and private sectors in new and creative ways with a sense of seriousness, national purpose and the urgency the situation merits. —Hon. Alexander Karsner, Assistant Secretary for Energy Efficiency and Renewable Energy, speech to the 2006 Power-Gen Renewable Energy and Fuels conference

- But efficiency advocates say the potential value of reduced consumption is far higher. Quite apart from slowing the increase in global warming gases, they say that because natural gas prices are generally determined by supply and demand in North America, cutting demand for it in American homes could cut price. —*New York Times*

- There is no cheaper, cleaner power than power you don't have to produce. —Gary Zarker, former superintendent, Seattle City Light

- Since achieving efficient use costs less than the fuel and electricity saved, the problem of climate change can be solved at a profit, rather than at a cost. And since making electricity-saving technologies needs about 10,000 times less capital than generating more electricity, the power sector—now a black hole for a quarter of the world's development capital—could become a net exporter of capital to meet other development needs. —Amory B. Lovins, "Small is Powerful"

- Energy efficiency is a well established, proven technique for both reducing carbon emissions and saving money. —*Financial Times*

- In this environment, the only near-term policy solution is to moderate demand through energy efficiency and conservation. Experience has shown that aggressive efficiency and conservation efforts can reduce demand growth and allow energy supply markets to catch up. And efficiency is not just a stop-gap solution—it must be sustained for the long term, because energy suppliers face continuing challenges in trying to bring increasing energy quantities to market fast enough. Constraints on capital and materials, environmental concerns, regulatory approvals, and other factors will continue to limit the rate at which supply infrastructure can be deployed. —ACEEE news release, "ENERGY EFFICIENCY CAN LOOSEN AMERICA'S ENERGY STRAITJACKET"

- Haste makes waste. —English proverb

- Waste not, want not. —English proverb

- Waste is too expensive; it's cheaper to do the right thing. —Paul Hawken, author and environmentalist.

- Our greatest national energy resource is the energy we currently waste. —Former Energy Secretary Spencer Abraham

- Present annual world energy consumption is about equal to the annihilation energy of 4 tons of matter. —Barney Oliver

- It's cheaper to save fuel than burn it. —Amory B. Lovins

- The watchwords to meet the challenges and uncertainties of the 21st century are efficiency and innovation. —Honorable John H. Dalton

Blackouts

- One of the blackout incidents that we had—we had a fatality accident in [city name] that involved a 4-year-old girl. We did not want to see those types of things happening. —City Energy Manager in a report to the California Energy Commission

- We had many branches that were caught up in the outages where we had to literally shut our doors. Our security systems are electronic, we can't run them without electricity. —Credit Union Facility Manger in a report to the California Energy Commission

- Three reported deaths have been tied to the outage. New York Mayor Michael Bloomberg said at least one person died as a result of the blackout, and at least one firefighter was injured. In Canada, Ottawa Director of Emergency Services Tony Dimanti said a 15-year-old died from injuries suffered in a fire, and another person died after being hit by a car during an altercation. —Greg Botelho, "Power returns to most areas hit by blackout"

- August 14th Blackout caused $4.5 to 10 Billion in lost economic activity- three deaths—Jimmy Glotfelty, U.S. Department of Energy

- When the lights go out, modern life as we know it grinds to a sudden halt… communications fail, water systems shut down, factory work is disrupted, food spoils, businesses lose money. —Spencer Abraham

Standby Power

Most of the research on standby power has been in the residential segment.

- Electrical appliances used in homes and offices consume some energy when they are left on standby mode or even switched off. The typical electricity loss for an appliance can range from as little as 1 W to as high as 30 W. This loss and the associated cost are not high enough to attract the attention of the consumer. But when such power losses of all home and office appliance are aggregated at the level of a country, the amount becomes significant and cannot be ignored. According to an estimate of the international Energy Agency, the total standby power demand of the residential sector in industrialized countries amounts to 15 GW. —Brahmanand Mohanty, "Standby Power Losses in Household Electrical Appliances and Office Equipment"

- Strange though it seems, a typical microwave oven consumes more electricity powering its digital clock than it does heating food. For while heating food requires more than 100 times as much power as running the clock, most microwave ovens stand idle—in "standby" mode—more than 99% of the time. And they are not alone: many other devices, such as televisions, DVD players, stereos and computers also spend much of their lives in standby mode, collectively consuming a huge amount of energy. —"pulling the plug on standby power," *The Economist*

- Plug home electronics such as TVs and DVD players into power strips. Then turn the strips off when the equipment is not being used. —Energy Secretary Samuel Bodman

- Many idle electronics—TVs, VCRs, DVD and CD players, cordless phones, microwaves—use energy even when switched off to keep display clocks lit and memory chips and remote controls working. Nationally, these energy "vampires" use 5 percent of our domestic energy and cost consumers more than $3 billion annually. —Lawrence Berkeley National Laboratory

- Hundreds of thousands of tonnes of carbon dioxide are needlessly produced every year by computers, digital set top boxes, chargers and many other products left on standby mode. —UK Environment Minister Elliot Morley

- The United States, for example, uses about 1,000 megawatts continuously (the output of one Chernobyl-sized power station) to run television sets that are turned off. Adding VCRs' and other household devices' standby loads roughly quintuples this waste. —Amory B. Lovins

Sustainability

- Part of the bargain, the social contract which allows companies to be as large as they are, is that they become engaged in the challenges the world faces, rather than dismissing them as someone else's problem. —John Manzoni Chief Executive, Refining & Marketing, BP

- We are committed to creating economic value, but we are not indifferent to how we do it. —Idar Kreutzer CEO, Storebrand

- CEOs could point out that profits should not be seen as an end in themselves, but rather as a signal from society that their company is succeeding in its mission of providing something people want—and doing it in a way that uses resources efficiently relative to other possible uses. —Ian Davis, Worldwide Managing Director, McKinsey & Company

- Businesses should see investment in environmental or social issues as a form of risk management because of the positive signals it sends to the public and to investors. One direct business benefit of good governance is that it is rewarded in the markets by lower discount rates. —Andrew Brandler, Chief Executive Officer, CLP

Availability of Oil

- ...The real problem is not scarcity but concentration. About two-thirds of the world's proven oil reserves lie in the hands of just five Persian Gulf countries. As the market share of those regimes soars, so too will the chance of disruption, embargo or worse. "The real trouble with oil." —*The Economist*

- New energy discoveries are mainly occurring in places where resources are difficult to extract physically, economically, and politically. —American Associates of Petroleum Geologists

- The Stone Age did not end for lack of stone, but the oil age may when the world runs out of oil. —*The Economist*

- ... if we improved the nation's car fleet by 3 miles per gallon, we could also eliminate our imports of Gulf oil. The fleet is now at 18.3. Many manufacturers have developed prototypes ranging from 65 mpg to about twice that. —Amory B. Lovins

- Peak oil is an emerging reality. With production already declining in all but a few major oil regions, an energy shortfall is inevitable. As demand for oil continues to grow, this shortfall can only mean disappointments for those around the world who aspire to live more like Americans, consuming their body weight in oil every week (150 pounds on average). —Francis de Winter and Ronald B. Swenson

- Half of all road transport fuel is burnt in built-up areas, despite the fact that half of all the journeys in such areas are less than five kilometers in length. —European Commission

- The world consumes 84 million barrels of oil a day. —Chevron advertisement

- The world consumes two barrels of oil for every barrel discovered.

So is this something you should be worried about?

The fact is, the world has been finding less oil than it's been using for twenty years now. Not only has demand been soaring, but the oil we've been finding is coming from places that are tough to reach. At the same time, more of this newly discovered oil is of the type that requires a greater investment to refine. And because demand for this precious resource will grow, according to some, by over 40% by 2025, fueling the world's growing economic prosperity will take a lot more energy from every possible source.

The energy industry needs to get more from existing fields while continuing to search for new reserves. Automakers must continue to improve fuel efficiency and perfect hybrid vehicles. Technological improvements are needed so that wind, solar and hydrogen can be more viable parts of the energy equation. Governments need to create energy policies that promote economically and environmentally sound development. Consumers must demand, and be willing to pay for, some of these solutions, while practicing conservation efforts of their own.

Inaction is not an option. But if everyone works together, we can balance this equation. We're taking some of the steps needed to get started, but we need your help to get the rest of the way.

—Chevron advertisement

- It took us 125 years to use the first trillion barrels of oil. We'll use the next trillion in 30.

So why should you care?

Energy will be one of the defining issues of this century. One thing is clear: the era of easy oil is over. What we all do next will determine how well we meet the energy needs of the entire world in this century and beyond.

Demand is soaring like never before. As populations grow and economies take off, millions in the developing world are enjoying the benefits of a lifestyle that requires increasing amounts of energy. In fact, some say that in 20 years the world will consume 40% more oil than it does today. At the same time, many of the world's oil and gas fields are maturing. And new energy discoveries are mainly occurring in places where resources are difficult to extract, physically, economically and even politically. When growing demand meets tighter supplies, the result is more competition for the same resources.

We can wait until a crisis forces us to do something. Or we can commit to working together, and start by asking the tough questions: How do we meet the energy needs of the developing world and those of industrialized nations? What role will renewables and alternative energies play? What is the best way to protect our environment? How do we accelerate our conservation efforts? Whatever actions we take, we must look not just to next year, but to the next 50 years.

—Chevron advertisement

- Two-thirds of the oil used in the United States goes for transportation. Passenger vehicles alone account for 40 percent of oil use, partly because the fuel economy of new cars and trucks is at a two-decade low. —Union of Concerned Scientists

- It is estimated that 50 percent to 80 percent of the tires rolling on U.S. roads are under inflated. Driving with tires that are under inflated increases "rolling resistance," wasting up to 5% percent of a car's fuel. We could save up to 2 billion gallons of gasoline annually simply by properly inflating our tires. —U.S. Navy

- Americans driving SUVs can expect to pay $180 more for gas in 2004 than they did in 2003, and passenger car drivers will pay $144 more. But hybrid electric car drivers will only pay between $50 and $67 more for gas in 2004 than they did in 2003. —Alliance to Save Energy

- If all the cars in the United States were equipped with the most efficient tires possible, the fuel savings would equal 400,000 barrels of oil per day. —U.S. Navy

- Improvements in automobile efficiency since 1973 are saving consumers $151 billion in 2004 alone—more than twice as much as the federal government spends each year on education. —Alliance to Save Energy

- In 2004, SUV drivers will spend about $1,225 on fuel, while passenger car drivers will spend only $976. Hybrid electric car drivers will spend between $350 and $450. —Alliance to Save Energy

- The difference between a car that gets 20 MPG (miles per gallon) and one that gets 30 MPG amounts to $1,800 over 5 years, assuming gas costs $1.80 per gallon and one drives 12,000 miles a year. —Alliance to Save Energy

- Transportation accounts for more than 67 percent of the more than 56 percent of its oil supply, and imports are expected to reach 70 percent over the next two decades. —U.S. Department of Energy

- The United States consumes almost 9 million barrels of gasoline daily—43 percent of total global daily gasoline consumption. —Alliance to Save Energy

- By the end of 2005, the number of hybrid vehicles on the road will more than triple. According to Department of Energy projections, by the end of this decade, 750,000 hybrid vehicles will be sold annually—that means one in every 23 passenger vehicles sold will be a hybrid electric. —Alliance to Save Energy

- Where gas or oil boilers are used for heating, an annual tune-up can avoid 5 percent of energy waste and can identify small problems before they become big problems—Energy Star

- Environmentally friendly cars will soon cease to be an option... they will become a necessity. —Fujio Cho, CEO Toyota Motor Corp

Recycling

Recycling will not reduce your energy bill, but it will save energy. If the goal of your organization includes a general mandate to save energy (verses energy costs), often the energy saved by recycling paper, cans, and other items can be counted towards the goal. The savings are real and tangible.

- You could watch more than two and half hours of television programming with the energy saved from recycling one aluminum can. —Bottlesandcans.com

- Recycled aluminum saves 95% energy vs. virgin aluminum; recycling of one aluminum can saves enough energy to run a TV for 3 hours. —Reynolds Metal Company

- Recycled aluminum reduces pollution by 95%—Reynolds Metal Co.
- Recycled glass saves 50% energy vs. virgin glass —Center for Ecological Technology
- Recycled glass generates 20% less air pollution and 50% less water pollution —NASA
- Recycled paper saves 60% energy vs. virgin paper —Center for Ecological Technology
- Recycled paper generates 95% less air pollution: each ton saves 60 lbs. of air pollution —Center for Ecological Technology
- Recycling of each ton of paper saves 17 trees and 7000 gallons of water —EPA
- Last year 54 billion cans were recycled saving energy equivalent to 15 million barrels of crude oil—America's entire gas consumption for one day. —Earth911
- According to the EPA, recycling a pound of PET saves approximately 12,000 Btus. —Earth911
- Recycled aluminum is identical to smelted aluminum, except for one thing: it takes only 1/20 of the energy to make it. Less energy means reduced greenhouse emissions. And like few other materials in the recycling chain, aluminum recycles over and over again. —Alcoa
- For each pound of aluminum recovered, Americans save the energy resources to generate about 7.5 kilowatt-hours of electricity. —Cancentral.com
- Recycling 1 ton of aluminum saves the equivalent in energy of 2,350 gallons of gasoline. —Cancentral.com

- Recycling one glass container saves enough energy to light a 100 watt lamp for 4 hours —EPA

- The aluminum can recycling process saves 95 percent of the energy needed to produce aluminum from bauxite ore, as well as natural resources, according to the Aluminum Association. Making a ton of aluminum cans from virgin ore, or bauxite, uses 229 Btus of energy. In contrast, producing cans from recycled aluminum uses only 8 Btus of energy per can. —EPA

- Saving material also saves the energy needed to produce, process, transport, and dispose of them. —Amory B. Lovins

- The energy saved from recycling one glass bottle will light a 100-watt bulb for four hours. —U.S. Navy

- Last year [2005], the 12 billion bottles and cans recycled by Californians saved the equivalent of enough energy to power up to 522,000 homes. —California Department of Conservation.

- Disposable (throwaway) bottles consume three times as much energy as reusable, returnable bottles. —U.S. Navy

- Producing aluminum from recycled aluminum consumes 90 percent less energy than producing it from raw materials and generates 95 percent less air pollution. —U.S. Navy

- Source reduction is to garbage what preventive medicine is to health. —William Rathje

Water Conservation

A large portion of the potable water bill, often 70% or more, is attributable to energy costs. Electrical energy is required to pump and move the water. Companies may choose to track the energy saved though water conservation activities and apply the same saving mythology.

- The biggest user of electricity in California is pumping water over the mountains. And therefore, water efficiency is a very good way to save electricity. —Amory B. Lovins

- Government cannot close its eyes to the pollution of waters, to the erosion of soil, to the slashing of forests any more than it can close its eyes to the need for slum clearance and schools. —Franklin D. Roosevelt
- Our world is increasingly burdened by the long-term risks associated with toxic chemicals. —Chris Bright
- Water and air, the two essential fluids on which all life depends, have become global garbage cans. —Jacques Cousteau
- A dripping faucet can waste up to 20 gallons of water a day. —U.S. Navy
- A leaking toilet can waste up to 200 gallons of water a day. —U.S. Navy
- A leaking faucet loses 796 liters (175 gallons) of water every month at one drop per second, wasting not only water but the energy used to heat it. —Peninsulas Health Care Corporation
- Just as energy planners have discovered that it is often cheaper to save energy—for instance, by investing in home insulation and compact fluorescent lights-than to build more power plants, so water planner are realizing that an assortment of water efficiency measures can yield permanent savings and thereby delay or avert the need for expensive new damn and reservoirs, groundwater wells, and treatment plants. —Worldwatch Institute's State of the World

LINKS TO OTHER RESOURCES

Energy Awareness Programs

- *EnergyShed.org* includes updated links to other awareness programs

- This BC hydro site provides many free energy awareness tools to get you started such as sample letters from management, a sample quiz, and posters you can print. *http://www.bchydro.bc.ca/business/investigate/investigate882.html*

- If you live in the UK Carbon Trust is a great resource. *http://www.thecarbontrust.co.uk/carbontrust/*

- California Flex Your Power has numerous tips and tools- *http://www.fypower.org/*

- Lawrence Berkeley National Laboratory site includes computational tools and spread sheets to calculate potential savings, along with sample posters and letters. *http://eetd.lbl.gov/bea/sf*

- If you live in New Zealand you can order posters from — *www.emprove.org.nz*

- Usually, "the cobbler's son has no shoes," however the U.S. Department of Energy leads by example with their internal energy awareness program "You Have the Power"—*http://www.eere.energy.gov/femp*

- U.S. EPA: The U.S. Environmental Protection Agency, Guidelines for Energy Management is a great resource. *www.Energystar.gov*

- U.S. Department of Defense (DON), The U.S. Department of Navy has good information at *https://energy.navy.mil/awareness/tools/tools_1.html*

- The University of Buffalo saved $10,000 in energy costs in a single day by reducing unnecessary energy consumption in their offices, labs and workspaces. *http://www.buffalo.edu/youhavethepower/yhtp.html*

- Your local university may have an awareness site, some examples are the "Conserve" site at the University of California at San Diego, http://conserve.ucsd.edu/Energy/Lab.htm, and Harvard's Green Campus Initiative, *http://www.greencampus.harvard.edu/cerp/*

Energy Efficiency

- Alliance to Save Energy—http://www.ase.orh
- American Council for an Energy Efficient Economy—http://aceee.org
- Association of Energy Engineers—http://www.aeecenter.org
- Energy Star—http://energystar.gov
- US Department of Energy—http://www.oit.doe.gov

Environmental Impact of Energy Consumption

- Airhead—http://www.airhead.org
- American Lung Association—http://www.lungusa.org
- BBC weather center—http://www.bbc.co.uk/climate
- Climate Analysis Indicators Tool—http://cait.wri.org/
- Intergovernmental panel on Climate change—http://ipcc.ch
- Natural Resources Defense Council—http://www.nrdc.org
- Pew Center for Global Climate Change—http://www.pewclimate.org
- U.S. Environmental Protection Agency—http://www.epa.gov
- U.S. Climate Technology Cooperation web —http://www.usctgateway.net (http://www.usctgatrwayonet/tool/

- U.S. Public Interest Research Group (U.S. PIRG)—http://www.usprig.org
- World Health Organization—http://www.who.int
- World Watch Institute—http://www.worldwatch.org

Non-technical Energy Awareness Glossary

A

Acid Rain—A broad term used to describe acidic (low pH) rain, fog, or snow due to the emission from fossil fuel burning power plants.

Active Power—The rate at which electrical energy is transferred, produced, or used. Sometimes called real power to differentiate it from reactive power. It is measured in watts (W), kilowatts (kW) megawatts (MW), etc.

Adjustable Speed Drive—See variable speed drive.

Albedo—The fraction or ratio of light that is reflected by a surface. Should global warming melt substantial amounts of ice on the planet, less light (and heat) will be reflected back into space. The extra heat will be absorbed by the land and sea and raise the earth's air temperature.

Alternative Fuels—Vehicle fuels other than standard gasoline such as ethanol, natural gas, propane, hydrogen, biodiesel, methanol, and electricity. Typically, alternative fuels have less harmful emissions and can be derived domestically from renewable resources

Ampere (A)—A unit of measure for electrical current.

Anthropogenic—Caused or originated by human activity. For example, anthropogenic atmospheric carbon dioxide is the amount of atmospheric carbon dioxide that is man made.

Apparent Power—The product of the voltage and the current often expressed in kilovolt-amperes (kVA).

Average Demand—The power output of an electrical system over any interval of time, as determined by the total number of kilowatt-hours divided by the units of time in the interval. See Demand.

B

Ballast—A device necessary to start and maintain the required energy flow in a fluorescent lamp.

Barrel—A unit of volume equal to 42 U.S. gallons.

Basal Metabolism (Energy Management)—The amount of heat given off by a person at rest; approximately 300—600 Btu per hour (Btu/h). The number of people in a building influences the energy required to cool the building. For reference, a single candle flame produces 850 Btu/hour.

Baseline—Reference data used for before and after comparisons

Baseload—The minimum amount of power that a utility must supply to its customers, or the minimum amount of power required by a customer.

Benchmarking (Energy Management)—The process of comparing a value to fixed value for comparison. For example, a company may decide to improve energy performance by 10% versus the prior year. The prior year usage then becomes the benchmark for which future energy is evaluated. Common benchmarked values compared to past performance, industry averages, and best in class values.

Blackout—Complete loss of electrical power.

British Thermal Unit (Btu)—A unit of energy, typically used to measure heat. The amount of heat required to raise the temperature of one pound of water one degree Fahrenheit. One Btu is equal to approximately 252 calories, 1055 joules, 0.0002928 kilowatt-hours. One ton of coal contains, on average, 28,000,000 Btu.

Brownout—A voltage drop in response to a shortage of power relative to end use demand. A brownout will cause a dimming of lights, effect the functioning of equipment, and may cause some devices to become inoperable.

Btu—The abbreviation for British Thermal Unit(s).

C

Carbon Dioxide (CO_2)—A colorless, odorless, nonpoisonous, gas that is present in the atmosphere. It is the principle gaseous product from the combustion of fossil fuels such as natural gas, oil, and coal. Electrical power generation through fossil fuel combustion produces significant quantities of Carbon Dioxide.

Carbon Dioxide Equivalent (CDE or CO_2e)—represents the comparable weight of carbon dioxide emitted into the atmosphere that produces the same global warming potential as a given weight of another gas. It is used to compare the emissions from other greenhouse gases based upon their global warming potential. Carbon dioxide equivalents are often expressed as million metric tons of carbon dioxide equivalents (MMTCDE).

Carbon Dioxide Capture and Storage—An alternative to emitting carbon dioxide into the air is an option of capturing and storing it in underground locations such as abandoned mines.

Chiller—A device for removing heat from a gas or liquid for air conditioning and cooling.

Climate—The general weather conditions of a geographic region.

Climate Change—The energy management definition describes the change in the climate as a result of fossil fuel combustion for energy consumption.

Coincident Demand—The kW demand reading of a consumer that occurs at the time of a power supplier's peak system demand.

Coincident Power Factor—The power factor at the time of a peak demand, or power supplier's peak system demand.

Combined Heat and Power (CHP)—Power plants designed to produce both heat and electricity from a single heat source.

Commissioning (Energy Management)—The process by which a building's energy intensive systems are set-up and approved for use. Ideally, this involves more than simply installing the equipment and flipping the switch, but rather a systematic and continual process to ensure the system is optimized for occupant or process, and energy efficiency.

Conservation (Energy Management)—To eliminate or reduce the consumption of energy.

Cooling Degree Day(s) (CCD)—See degree day

Cooling Tower—A device used to cool power plant water by evaporation.

Cryosphere—Parts of planet that are seasonally or perennially frozen or covered by snow and ice. Changes in the volume of cryosphere impacts the climate because it reflects more incoming solar radiation.

D

Daylighting—The use of sunlight to provide light in building interiors.

Degree-day—A unit measure of heating and cooling by measuring the extent to which the outdoor temperature falls above or below an assumed base. The base is normally taken as 65 °F or 18 °C. One degree-day is counted for each degree below for heating or over for cooling of the assumed base, for each day on which it occurs. For example, a temperature of 70°F is 5 cooling degree days, an average temperature of 70°F for three days is 15 cooling degree days.

Delamping—Removing unnecessary lamps (bulbs) from multi-lamp fixtures. For example, a florescent fixture many hold four lamps, upon examination it may be determined that two or three lamps are sufficient to properly illuminate an area and the excessive bulbs are removed, thereby reducing energy costs. Be sure to verify the lighting levels and check with the ballast manufacturer before delamping.

Demand—The rate at which electrical energy is delivered to devices at a given instant or averaged over a specified period of time, typically 15 minutes. Billed in kilowatt (kW) or equivalent units. See Kilowatt.

Demand-side Management (DSM)—Managing the consumption of energy in conjunction with the power provider to optimize available generation resources.

Dimmer—A light control device that allows light levels to be manually or automatically adjusted. A light dimmer can save energy by reducing the amount of energy delivered to the lamp.

Distributed Generation—Local or on-site power generation located near the point where the power is used.

Duty Cycle—The proportion of time during which a device is operated. The duty cycle can be expressed as a ratio or as a percentage.

E

Economizer—An air economizer system will bring in outside air for building cooling when the outside temperature is cool enough and the building requires cooling. It is more economical to shut off the compressor and bring in cool outside air to satisfy the cooling needs of the building when the outside temperature and humidity meet the necessary requirements.

Electrical Energy—The energy delivered through the moving of electrons, usually measured in kilowatt-hours (kWh). See kilowatt hours.

Electrical System Losses—The amount of useful and billable energy lost during generation, transmission, and distribution of electricity. Much of the losses are given off as heat.

Electric Power Plant—A facility or device that produces electrical energy.

Emissions (Energy Management)—Release of gases to the atmosphere during generation of energy. Greenhouse gas emissions are typically measured in tons per time period.

End Use—The purpose for, device, or place where useful energy is consumed.

Energy—The capacity for doing work. Electrical energy is usually measured in kilowatt-hours (kWh), while heat energy is usually measured in British thermal units (Btu). One kWh is 1,000 Watts and is equal to 3,413 Btu

Energy Audit—A survey and process that finds ways to use less energy.

Energy Awareness—Refers to an individual's knowledge and observance of energy issues.

Energy Awareness Program—A comprehensive program designed to raise energy awareness and make individuals accountable for energy consumption.

Energy Efficiency—The ability to produce the same amount of heating, lighting, manufacturing production, or other energy consuming activity with less energy. That is, getting more use out of the energy we already generate. Measures of energy efficiency can be the same as energy intensity (energy dollars)/(square foot), (energy dollars)/(GDP), (energy dollar/unit of production).

Energy Intensity—The amount of energy use per unit output, such as energy consumption per square foot of office space (kWh/ft^2).

Energy Management—The administration, handling, or process of managing the use, cost, and reliability of power derived from sources such as fossil fuel, electricity and solar radiation.

Energy Management and Control System (EMCS)—The general term for an automatic system used for controlling HVAC, lighting or power equipment in a facility. Typically, computer based including either electrical or pneumatic controls.

Energy Service Company (ESCO)—A company that specializes in undertaking energy efficiency measures under a contractual arrangement whereby the ESCO shares the value of energy savings with their customer.

Energy Source—A substance that can be transformed to supply energy. Examples include petroleum, coal, natural gas, wind, sunlight, and water movement.

Energy Star—A voluntary government-backed program designed to help businesses and individuals protect the environment through superior energy efficiency. The primary goal of ENERGY

STAR is to prevent air pollution by expanding the market for energy-efficient products. The ENERGY STAR label is a mechanism that allows consumers to easily identify energy efficient products that save energy and cost less to operate. The label removes information barriers that affect purchasing decisions. By raising environmental awareness, ENERGY STAR stimulates demand for high-efficiency products.

Entropy—A measure of the unavailable or unusable energy in a system, usually heat, that cannot be converted to another form. The entropy of the universe is always increasing.

ExaJoule—1 Joule followed by 18 zeros or 278,000 GWh. See Joule.

F

Final Energy—The energy we actually use in our homes, businesses, and cars.

Fossil Fuels—Energy sources formed in the ground from the remains of plants and animals. Common fossil fuels are oil, natural gas, and coal.

Fuel—Any material that can be burned to make useful energy.

Fugitive Emissions—Unintended leaks of gas from the transmission, transportation, or processing of fossil fuels.

G

GDP—An acronym for Gross Domestic Product, a measure of the size of an economy.

Generator—A device that converts mechanical energy to electrical energy.

GigaJoule (GJ)—One billion Joules (a joule followed by 9 zeros). 1 GJ = 277.8 kWh.

Gigawatt (GW)—1 billion Watts (W), 1 million kilowatts (kW), or 1,000 megawatts (MW).

Gigawatt-hour (GWh)—1 billion watt hours, 1 million kilowatt-hours (kWh), or 1 thousand megawatt hours (MWh).

Global Warming—A term used to describe the increase in average global temperatures due to the greenhouse effect. The term is most often used to refer to the warming some scientists predict will occur as the result of emissions from burning fossil fuels, or the release of other greenhouse gas emissions.

Greenhouse Effect—The result of atmospheric gases trapping radiant infrared energy keeping the earth's surface warmer than it would otherwise be.

Greenhouse Gases—Gases in the earth's atmosphere such as water vapor, carbon dioxide, nitrous oxide, and methane, that are transparent to solar radiation, but opaque to long wave radiation thus allowing the atmosphere to retain heat. Greenhouse gases occur through both natural and human induced processes. The main greenhouse gases are: carbon dioxide, methane, nitrous oxide, hydrofluorocarbons, perfluorocarbons, sulfur and hexafluoride.

Green Power—Energy produced from clean renewable energy resources.

H

Heating Degree Day(s) (HDD)—See degree days.

Heating, Ventilation, and Air-Conditioning (HVAC) System—All the components of the system used to condition interior air of a building.

Hibernation Mode—See stand-by power.

High-intensity Discharge Lamp—High-intensity discharge (HID) lamps produce a large quantity of light in a relatively small package. The lamp consists of a sealed arc tube inside a glass envelope, or outer jacket. The inner arc tube is filled with elements that emit light when ionized. Like fluorescent lights a ballast is required start and regulate current during operation

Horsepower (hp)—A unit of power. 1 hp is equal to 0.746 kilowatts or 2,545 Btu per hour.

I

Incandescent Light—The most common type of light used in residential settings. These lamps use an electrically heated filament to produce light in a vacuum or inert gas-filled bulb. The vacuum or gas stops the filament from burning. If the glass is broken oxygen from the atmosphere will immediately burn an energized filament. Incandescent lights are inexpensive and provide a pleasing color of light, but use more energy than most other lights. Many energy managers refer to incandescent lights as small heaters as the majority of energy consumed is given off as heat.

Interruptible Load—Energy end use loads such as lights and motors that can be shut off or disconnected at the energy supplier's discretion or as determined by a contractual agreement.

J

Joule—A unit of energy or work. The energy produced by a force of one Newton operating through a distance of one meter. Since a Joule is a small amount of energy it is typically expressed as a GigaJoule (one billion Joules) or an ExaJoule (a Joule with 18 zeros). 1 Joule per second equals 1 Watt or 0.737 foot-pounds; 1 Btu equals 1,055 Joules.

K

Key Performance Indicator (KPI)—A specific measure of performance that we want to know. Examples of energy management KPIs are: Cost of energy consumption per square foot ($/sqft), electrical energy consumption per widget (kWh/widget), chiller energy per output (KW/ton), or total energy per square foot (kBtu/sqft) which is commonly called "energy index." A variation is Performance Measures (PM), which is the information we have and can report whereas KPIs are the information we ideally want.

Kilowatt (kW)—The basic unit of electrical power equal to one thousand watts, or the energy consumption at a rate of 1000 Joules per second. The k in kWh represents the number 1,000. 746 kW equals 1 horsepower. See Watt

Kilowatt-hour (kWh)—The kilowatt-hour is the basic unit in which residential electrical energy is bought and sold. It is often referred to as energy consumption, or some times just electrical energy. Think of a kWh as you would an odometer on a car; it's a cumulative sum. 1kWh = 3,412 Btu

L

Leaking Electricity—Standby power

Lighting Sweep—A strategy that turns off banks of lights at various intervals. Typically used during unoccupied hours such as at night. A typical scenario will turn off groups of lights at one-hour intervals through out the night. That way, if light were left on they will be turned off until overridden. Some system will flash first to give occupants a chance to override the sweep in advance.

Line Loss—Electrical energy lost because of the transmission or consumption of electrical energy. Much of the loss turns into heat. Line loss averages around 10%.

Load—The power required to run a circuit, system, light, air conditioner, or an entire building.

Load Profile—A graph showing power (kW) supplied plotted against time to illustrate the variance in a load in a specified time period.

Load Shedding—Turning off or reducing loads to limit peak demand.

Load Shifting—Moving loads from on-peak periods to off-peak periods.

M

Master Meter—A meter that records the consumption of electrical energy, gas, water, etc, and at only one point.

Megawatt (MW)—1 thousand kilowatt (kW), or 1 million watt (W).

Megawatt-hour (MWh)—1 thousand kilowatt-hours (kWh) or 1 million watt-hours (Wh). A megawatt hour is approximately the energy required to power 1,000 homes for one hour.

Motor—A machine that converts electrical energy into mechanical force and/or motion.

Mtoe—Megatons of oil equivalent. 1Mtoe = 11630 GWh.

N

Nonrenewable Fuels—Fuels we cannot replenish or create such as oil, natural gas, and coal.

O

Occupancy Sensor—A device that automatically turns room lights on when person's presence is detected and later turns lights off after the space is vacated.

OECD Countries—Countries that signed the Convention on the Organization for Economic Co-operation and Development.

They are Australia, Austria, Belgium, Canada, Czech Republic, Denmark, Finland, France, Germany, Greece, Hungary, Iceland, Ireland, Italy, Japan, Korea, Luxembourg, Mexico, Netherlands, New Zealand, Norway, Poland, Portugal, Slovak Republic, Spain, Sweden, Switzerland, Turkey, United Kingdom, and the United States.

Off-Peak Energy—A period of low energy consumption, as opposed to on-peak demand. See time-of-use.

Office Equipment—Includes (but not limited to) computers, monitors, computer speakers and accessories, printers, copiers, task lights, fax machines, scanners, multifunction devices, transformers, PDAs and cell phones.

On-Peak Energy—Energy supplied during periods of relatively high system demands as specified by the supplier. See time-of-use.

P

Part Per Million (PPM)—A measure of the parts of a substance contained in a million parts of another substance. Often used as a measure to express the amount of carbon dioxide or greenhouse gases in the atmosphere.

Peak Demand—The maximum kW energy demand in a specified time period, typically 15 minutes.

Peak Oil—Refers to the global maximum peak in oil production after which oil production begins to decline. There is uncertainty when this will happen, popular predictions range from 1977 to 2030, but it as yet to occur.

Peak Shaving—Typically peak shaving refers to starting on-site generators to reduce peak load, and thus differs from load shedding.

Percent Change—The change in a number over a specified time period. Calculated as the current value minus the previous value divided by the absolute value of the previous value. This new number is then multiplied by 100 and a percent sign is added. For example, if the prior consumption was 2,000 kWh per day and the new consumption is now 2,500 kWh per day. The percent change is (2,500-2,000)/2000 = .25 = 25% increase in daily kWh consumption.

Phantom Load—See stand-by power.

Plug Load—Energy used by devices connected to ordinary wall outlets.

Power—The rate at which energy is transferred, produced, or consumed. It is measured in horsepower, Btu per hour, or with electrical power it is measured in watts (W), kilowatts (kW) megawatts (MW), etc.

Power Factor (PF)—The ratio of actual power (W) to the power that is apparently being drawn from a power source (VA). See reactive power or description in Chapter 1.

Primary Energy—The total available energy from resources such as coal, oil, and natural gas assuming a 100% efficient use of those resources.

R

Ratchet—A peak demand that occurs only once a year, but used by utilities to set a minimum monthly charge on a go forward (or retroactive) basis. The reasoning behind a ratchet is that utilities need to provide the maximum required infrastructure year around even if the maximum usage is needed only once a year, so the utility may charge for the one time peak demand (ratchet) every month even though it was only hit once a year. Often, ratchets are reset once a year.

Reactive Power—The portion of electrical power that is used to create and sustain magnetic fields in alternating-current equipment. It is expressed in kilovars (kvar) or megavars (Mvar).

Real Power—See "Active Power."

Recycling—The action or process of converting materials into a new product.

Relamping—The replacement of a lamp with a new lamp. Since labor is an expressive component of the cost of changing lamps, commercial and industrial companies will often relamp or replace numerous functioning bulbs at one time. The term usually refers to the replacement of lamps with more energy efficient lamps.

Renewable Energy—Energy derived from resources that are regenerative or for all practical purposes cannot be depleted. Types of renewable energy resources include moving water, solar, and wind energy.

Retrofit—The process of modifying an existing building or process. A term used to distinguish from a green field project.

S

Sag—A short period of lower than normal voltage causing motor heating or equipment problems. Usually considered to be less than one minute in duration or it is considered an under voltage condition rather than a sag.

Setback/Setup—A strategy where thermostat setpoints are automatically changed during unoccupied hours to a reduced (setback) temperature during the winter and increased (setup) temperature during the summer.

Set-top Boxes—Typically any of the following devices: cable TV boxes, digital converters, internet access devices, video games, videophone boxes, cable modems, satellite boxes, wireless TV boxes, personal video recorders (VCR), DVD players, TiVo, Slingbox, and similar multifunction devices. In 2002, there were approximately 100 million settop box units in the U.S. with a projected growth to nearly 200 million units by the year 2010.

Standby Power—Typically refers to the power consumption of a device that is switched off or not performing its primary function. This may also be called "hibernation mode," "free-running power," "leaking electricity," "off-mode power," "phantom loads," "standby losses," "standby mode," "sleep mode," and "waiting electricity." Standby power has a surprisingly large impact on energy consumption and the global environment. In many new business computers System Standby (S3 sleep state) and Hibernate (S4) both enter a low power sleep state of 1 to 3 watts. System Standby can wake-up 2 to four times faster (5 to 10 seconds) by storing data in RAM rather than the hard drive as Hibernate mode does (Hibernate saving data on the hard drive better protects the data in the event of a power loss).

Sub-metering—The process of metering and monitoring each tenant, cost center, or individual loads. As opposed to Master Metered.

Surge—A short period of higher than normal voltage that may damage equipment.

T

Task Lighting—A light source specifically to light a task or work performed by a person or machine.

Terawatt (TW)—One trillion watts.

Terawatt-hour (TWh)—One trillion watt-hours.

Therm—A unit of the heat content of gas equal to 100,000 Btu.

Thermal Mass—Any material that stores heat.

Thermodynamics—A branch of physics that deals with energy (heat) and the conversion of various forms of energy. The first law of thermodynamics is that energy cannot be created or destroyed, but only changed from one form to another. This is relevant to global warming as the heat transferred from the sun should equal the heat radiated away or the earth's temperature will change.

Time-of-day Schedules—A time based program that dictates when equipment is turned on or off.

Time-of-use (TOU) Rates—A method of billing where the price of electrical energy varies during a particular block of time. Time-of-use rates are usually divided into two, three or four time blocks per twenty-four hour period and by season of the year. Time blocks are often called on-peak, mid-peak, or off-peak. An example may be where a kWh cost $0.14 cents from 8:00 A.M to 9:00 P.M (on-peak) and $0.08 the rest of the time (off-peak).

Ton—A U.S. unit of weight equal to 2,000 pounds (short ton). A British unit of weight equivalent to 2240 pounds (long ton). A unit of weight equal to 1,000 kg (metric ton). A unit of air cooling capacity equal to 12,000 Btu per hour.

Transformer—A device that changes alternating current electricity from one voltage to another.

V

Variable Frequency Drive (VFD)—An electronic device that controls the speed of motor-driven equipment such as fans and pumps. Motors typically consume less energy when operated at slower speeds with variable frequency drives. Also called adjustable frequency drives and variable speed drives.

Volt—A unit of electrical force equal to that amount of electromotive force that will cause a current of one ampere to flow through a resistance of one ohm.

Voltage—The amount of electromotive force, measured in volts, that exists between two points.

Volt-Ampere (VA)—A unit of electrical measurement equal to the product of a volt and an ampere. See Apparent Power.

W

Watt (W)—A measure of the rate of energy use sometimes called power or demand. The rate of energy transfer is equivalent to one ampere under an electrical pressure of one volt. Think of watts as the vehicle equivalent of a speedometer, it's an instantaneous reading. A 100 watt light bulb uses energy at the rate of 100 watts. One watt is equivalent to 1/746 horsepower, or one joule per second.

Watt-hour (Wh)—A unit of electrical energy. The consumption of one watt over the period of one hour. A kWh, or 1,000 watts is a common residential utility billing determinate. See kWh.

Work Station—Both the computer and monitor.

Y

Year to Date—The cumulative sum of each months value starting with January and ending with the current month of the data.

Z

Zone—An area within the interior space of a building, such as an individual room(s) to be cooled, heated, or ventilated. A zone has its own thermostat to control the flow of conditioned air into the space.

Index